BUILDI
WILLIAMSON
ENGINE

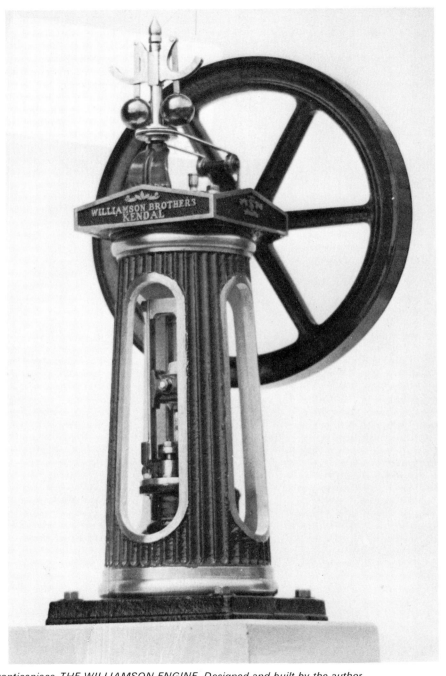

Fronticepiece-THE WILLIAMSON ENGINE. Designed and built by the author.

BUILDING THE
WILLIAMSON
ENGINE
Tubal Cain

Model & Allied Publications, Argus Books Limited

ISBN 0 85242 719 0

Model & Allied Publications
Argus Books Ltd.,
14 St. James Road, Watford,
Hertfordshire, England.

The text of this book is based on articles from
the *Model Engineer.*

Printed and bound in Great Britain by
Pindar Print Ltd., Scarborough.

CONTENTS

FOREWORD

The pseudonym 'Tubal Cain'—'an instructor of every artificer in brass and iron', conceals the identity of a retired professional engineer whose personal experience ranges from reciprocating steam engines to gas turbines, and much between. One of a family of engineers going back several generations, (and his son and grandson are following in his footsteps) he spent his boyhood wandering round iron and steel works before a training which involved not only working on most machine tools, but extensive experience on the shop floor, building and testing engines etc. He later became a leading designer and was given charge of all research and development before turning his attention to technical education. Here his experience ranged from teaching elementary craft studies to postgraduate degree work—an instructor of artificers indeed, though the range of engineering materials has widened a bit since Genesis!

Forty five years ago 'Tubal Cain' bought his first lathe, a $1\frac{5}{8}$ in. 'Adept', and has been making models ever since, not because he has to but because he likes to. Again, this is not exceptional, many have made a similar start. But 'Tubal Cain' has the unique gift of applying his considerable skill and expertise in engineering to his model making activities and to showing others how to do likewise. This is nowhere more apparent than in the design of his engines, the *Williamson Engine* and *Mary*, for example. They are really *designed*, just as a full size engine is designed, and not just put together from bits and pieces. Every small detail has a function and a purpose, and with 'Tubal Cain' they are always right for the type and age of the engine which he is designing in miniature.

In his instructions for machining, fitting and erecting he relies very much on his own past experience of, and preference for, the days when fitters really had to fit and not merely assemble. This too is not inappropriate to the amateur with limited skill and equipment since it ensures that with care and patience a workable engine will always be produced.

The readers of this book, and others which may follow, will find a great deal to interest them, whether it be in the elegance of the design, the method of machining, the fitting and erecting of the piece parts, or in the construction of engines which incorporate all that was best in the Age of Elegance.

D. H. Chaddock
C.B.E., M.Sc., C.Eng., F.I.Mech.E.

Chapter One

MAKING A START

The first vertical engines were built very much as were the contemporary beam engines, with the cylinder on the floor and the crankshaft supported by the walls of the engine-house. The cylinder was the heaviest part of the engine and the floor was the 'proper' place for it! After the introduction of the 'high pressure, engine around 1820, engines became much smaller, especially those without condensers, and the vertical type was made self-contained. A variety of forms was adopted but that incorporating an ornamental column was, perhaps, the most usual. The type of engine known today was introduced rather later, and was known as the 'Inverted Vertical' or 'Steam Hammer' engine, with crank at floor level and cylinder above.

The little engine shown in fig. 1 is, I think, one of the most attractive ever built. It was made by Williamson Brothers of Kendal, Westmorland, and shown at the International Exhibition of 1862. There it attracted not only a mention but also an illustration from the great D. K. Clark in his account of the exhibited machinery, published in 1864. The brothers Henry and William Williamson set up a small agricultural machinery business in the village of Stainton, Westmorland, in 1853, but moved to Kendal in 1856, occupying premises in one of the canal head warehouses. Their principal manufacture was agricultural machinery, but they also made domestic washing machines, mangles, and even lawn-mowers. The normal method of driving

threshing machines in those days was the 'horse gear' with one, two, or four horses, but Williamsons also offered water wheels, the then newly introduced 'Thompson' water turbine (a good many of these were sold to farmers in Westmorland) and steam engines. The latter ranged from little 4 h.p. verticals to engines as large as 16 in. bore, and they

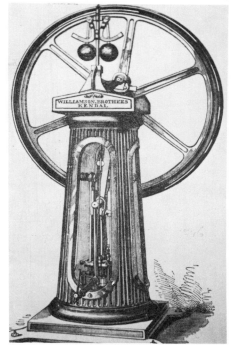

Fig 1 - The Williamson Columnar Engine. (From 'The Exhibited Machinery of 1862' by D. K. Clark. [London 1864].)

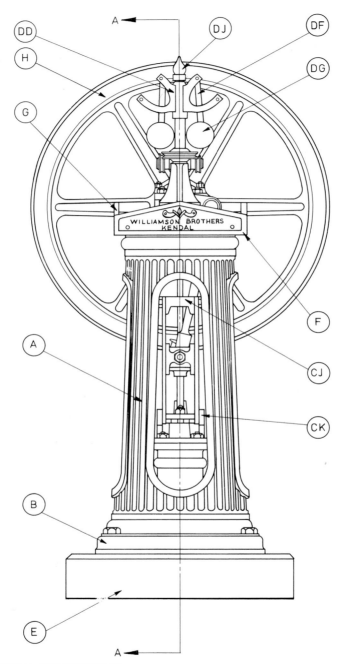

WILLIAMSON BROTHERS
KENDAL

Fig 2-GENERAL ARRANGEMENT.
(The part letters are used throughout the book.)

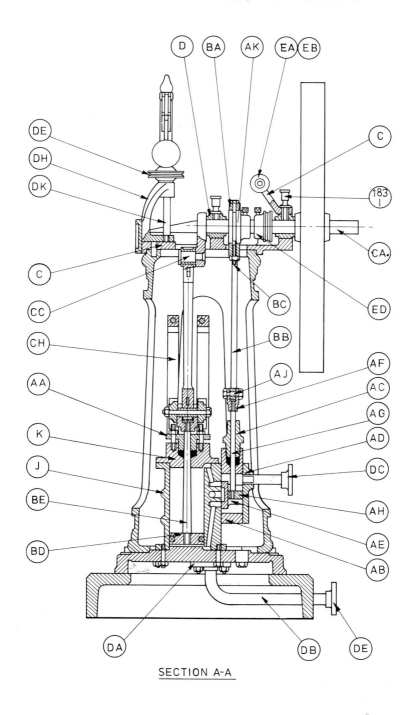

SECTION A-A

11

also listed a 'Portable' engine.

The prototype of this model is their 5 h.p. engine, of $6\frac{1}{2}$ in. bore, and is, in fact, the one exhibited in 1862. The stroke is not stated, but appears to have been about 14 in., the speed being perhaps 130 r.p.m. The price, including a simple boiler with usual fittings, was £90 delivered in Kendal. Incidentally, the works is still in being; the firm became Gilbert Gilkes & Co. making water turbines in 1881, and is now Gilbert Gilkes & Gordon Ltd., in the same line of business. It is from a catalogue in their possession that I have designed the model, to a scale of one-tenth full size.

The general arrangement (fig. 2) shows that I have had to take a few liberties with the design. The cylinder of the original was partly in the floor, with pipes in trenches. I have mounted the cylinder on the base, and provided a plinth. The top anchorage of the crosshead guides is different (on the original the chaps could get inside to do the fitting!) and the

governor does not control the engine speed; further, it is driven by belt, not bevel gears and return crank.

Readers who are not afraid of watch-making jobs may care to modify the design in this respect, but they will find that, as on the original, the linkage is quite complex!

Castings and all screws and materials are supplied as a 'package' by Messrs. Stuart Turner, with working drawings. (Fig. 3). I suggest that you treat yourself to a 6 BA plug tap if you haven't got one, as one set of holes must be tapped to the bottom. You will also need a 10 BA seconds tap for the eccentric, so order one of these too. Not supplied in the package is some Araldite adhesive, needed to attach the decorative top sideplates. How-ever this material is readily available but don't accept the so-called 'rapid' or quick-curing variety. Get the ordinary 2-tube pack in the blue box; the rapid works too fast for our purpose—and is a flaming nuisance anyway!

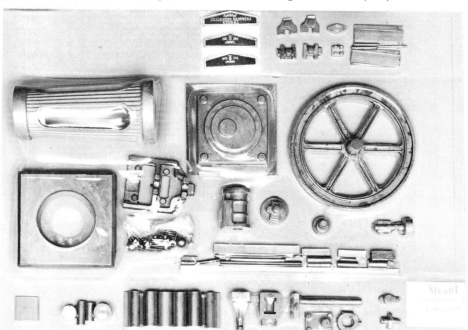

Fig 3-Castings and all materials are supplied as a package.

Fig 4-Flywheel end view of the engine.

Having checked all the parts from the schedule, I recommend a coat of cellulose primer to all iron castings, both to prevent rust if you don't use them immediately, and to keep oil out of the pores in manufacture. Run your rule over the castings and get used to the various parts by comparison with the drawings. Naturally you will trim any flashes or rough patches at this stage—get all the mucky work over first—and if any parts have a right and wrong way round, mark them with a felt pen. Which bits you start with is up to you, but it pays to machine all parts of the same material at once, especially if you use a 4-tool turret.

Chapter Two

ENGINE STRUCTURE

The photograph in fig. 4 gives a fair idea of how the base, Column, and entablature (top) fit together. The Column is the critical part, so machine this first.

Column, Part A fig. 5

This will come as an aluminium casting; the prototype shown in my illustrations is an iron one, used in developing the patterns for later casting in light alloy, and will differ slightly in appearance. However, the machining processes will be no different, though you will have rather less metal to take off than I did, and won't have to do as much work (if any at all) on the flutes. You will need to trim off the casting, and especially square off the ends if there is any flash there. I also recommend that if the flutes need any attention, or the edges of the apertures in the column, you deal with these before machining, when it is possible to hold the work in the vice without risk of damaging machined surfaces. Having done this, chuck the casting, large end outwards, in the 4-jaw and set the whole running true—work to the fluted part, and check both ends repeatedly until you have it as true as possible. Get a good grip with the chuck, but don't use excessive force, as you have a long overhang for the first cuts. Figure 6 shows the set-up. (Again, note this is the iron casting.)

Set up a good stiff boring bar and bore the hole in the column until you get a good clean bore. Into this fit a taper plug of hardwood (I use boxwood, but any fine grain wood will do, or some would use lead) and centre this with a large Slocomb. Bring up the tailstock but don't use too high a centre-pressure. As an experiment I tried this operation without out the tailstock, and no catastrophe followed! If you use too high a centre pressure you may disturb the chuck setting. Check carefully the dimension between top and bottom rim round the flutes. It should be $\frac{13}{16}$ in., but not to worry if it's $\frac{1}{32}$ in. out. But if it is out, make sure you will have the full $\frac{3}{8}$ in. available when you machine the face—it is more important that this be available for the rim and radius at the bottom than that the top be $\frac{7}{16}$ in. With the centre in place, do any rough machining necessary on the diameter and then face the end, leaving just a few thou. on.

Take out the plug and finish the bore to size; caliper dimension will do, as you can make the base spigot to fit. Finally, take a light skim off the face to ensure that this is really true to the bore. We now have a reference face and a true bore on which to base all future machining. You will appreciate that all the alignment of the working parts depends on this—the virtue of the column construction is that alignment is held once it's right; the defect is that if it's wrong, it stays wrong! We can now locate the column on a spigot whilst machining the top, and so ensure that this, too, will be true.

Find a piece of brass or steel about $2\frac{1}{2}$ in. diameter and long enough to sit back on the face of the 4-jaw and project

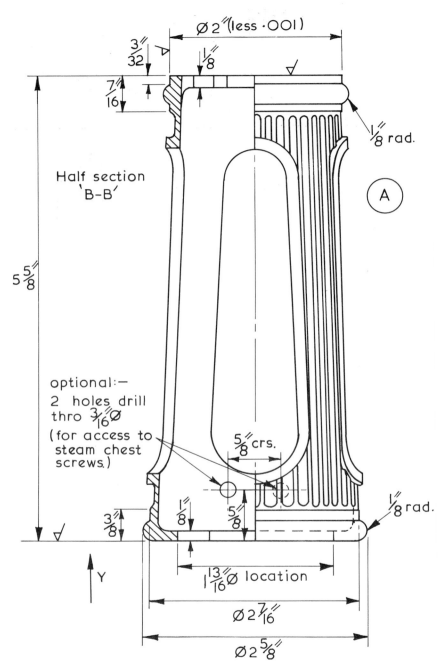

Ø 2″(less ·OO1)

$\frac{3}{32}$″

$\frac{1}{8}$″

$\frac{7}{16}$″

$\frac{1}{8}$″ rad.

Half section
'B–B'

Ⓐ

$5\frac{5}{8}$″

optional:–
2 holes drill
thro $\frac{3}{16}$″Ø
(for access to
steam chest
screws.)

$\frac{5}{8}$″ crs.

$\frac{3}{8}$″

$\frac{1}{8}$″

$\frac{5}{8}$″

$\frac{1}{8}$″ rad.

Y

$1\frac{13}{16}$″Ø location

Ø $2\frac{7}{16}$″

Ø $2\frac{5}{8}$″

Fig 5-The column and plinth.

16

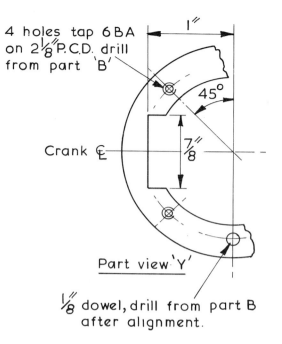

4 holes tap 6 BA
on $2\frac{1}{8}$" P.C.D. drill
from part 'B'

1"

45°

Crank ₵

$\frac{7}{8}$"

Part view 'Y'

$\frac{1}{8}$" dowel, drill from part B
after alignment.

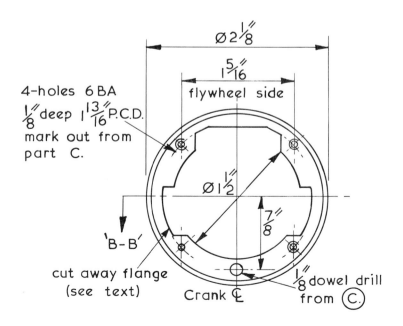

Ø$2\frac{1}{8}$"

$1\frac{5}{16}$"

flywheel side

4-holes 6 BA
$\frac{1}{8}$" deep $1\frac{13}{16}$" P.C.D.
mark out from
part C.

Ø$1\frac{1}{2}$"

$\frac{7}{8}$"

'B-B'

cut away flange
(see text)

Crank ₵

$\frac{1}{8}$" dowel drill
from Ⓒ

17

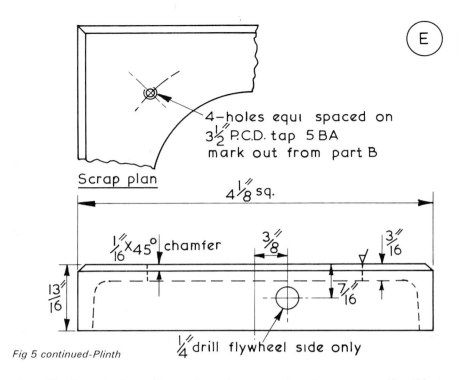

4–holes equi spaced on
$3\frac{1}{2}''$ P.C.D. tap 5 BA
mark out from part B

Scrap plan

$4\frac{1}{8}''$ sq.

$\frac{1}{16}'' \times 45°$ chamfer

$\frac{3}{8}''$

$\frac{3}{16}''$

$\frac{13}{16}''$

$\frac{7}{16}''$

$\frac{1}{4}''$ drill flywheel side only

Fig 5 continued-Plinth

about $\frac{5}{8}$ in. from the jaws. So you haven't got any? Go round to the bowls club and see if you can scrounge an old bowls wood that has split. These are made of lignum vitae, and this is admirable stuff for this sort of job; you will find that, though hard, it turns like cheese. (Might even convert you to wood-turning, at that!). Face one end of the chunk, and then reverse in the chuck and knock this face hard back onto the chuck body, with the chuck jaws tight. Turn the projecting end to $2\frac{1}{4}$ in. diameter, face the end, and form a spigot to fit the hole you have machined in the column base. Finally, drill a hole, say $\frac{7}{16}$ in. diameter, to pass a drawbar. Find a piece of rod (anything over $\frac{5}{16}$ in. will do) and screw both ends, long enough to go through the mandrel and project a couple of inches into the column.

Make a cross-piece to go through the apertures in the casting—see fig. 7 for the arrangement—but it must not project any more than is necessary. Use this to pull the casting hard back onto the spigot. Don't forget to use washers, a brass one where it beds on the tail of the mandrel. Note that it is vital that the stock on which the spigot is turned be hard up against the face of the chuck before you turn the spigot, or the draw-bar will upset all. You can now face the end of the casting, leaving a few thou. on and, if needed, do any roughing needed on the outside diameter. Fit a plug and use the tailstock centre if you wish. Now machine the 2 in. dia. spigot on the top. I know this says '2 in. less 1 thou.' and you haven't a 2 in. mike; never mind—it only means that the spigot should be a snug fit in its mating hole. Finally, finish the face to length—work as close as you can, but the cylinder clearance spaces are sufficient to allow you 'half a sixty-fourth' if that is the best you can do.

Next to finish the shapely curves. You can make a form tool if you wish, but I

Fig 6-First operation on the column.

Fig 7-Machining the top of the column.
(Note the heavy spigot-piece held in the chuck, and the draw-bar cross piece.)

Fig 8-A finished column.
(Note: This is the initial design. The present column is not bored out at the top.)

did it by hand turning. If you have no handrest, set up a piece of $\frac{3}{8}$in. or $\frac{1}{2}$in. square bar in the toolpost—the bigger the better so long as the tool can come down to centre-height. It must be as close to the job as you can get, hence the reason for the short cross-bar through the apertures of the column. If you have no hand tools, use a medium or large hand scraper. Sharpen it well, and hone the cutting edge to a high polish. It must be sharp and kept sharp. Run in slow back-gear till you get the feel of it, and then speed up so long as you get no chatter. Start with the top end—fit a plug and use the tailstock if needed—and bring the cutting edge up to the work with *very* slight pressure and the scraper flat on the top of the rest. It may only just cut, or only just not cut. If it wants to cut hard, the rest is too low. Adjust it. Then, to put on cut, apply a trifle more pressure and very slightly lift the handle—only a trifle. Keep your other hand holding the cutting edge down on the rest—hard. The ideal position of the rest is when the tool just cuts when it is flat on top of the rest, more when the

handle is raised a little, and stops if it is lowered. As soon as you have the feel of it, work round the curve, from large diameter to small, till it looks right, well balanced on either side of the major diameter. There is a little radius where the bulb meets the parallel part—this can be formed afterwards with the sliderest.

Now for the other end. Here the chuck may get in the way, and you will have to use an angle tool. If you haven't one, make it from a flat file—one that is about $\frac{1}{8}$in. thick. Grind off all the teeth, and then a bit more, and form the end to a 45 degree angle. On no account allow the end of the tool to colour—use your water pot frequently in the process. Polish out the grinding marks with a smooth stone, and then heat the shank until the business end turns pale straw. (Files are not tempered.) Fit a really good handle and make sure it has a good grip of the tang—right on. Having sharpened and honed the cutting edge, proceed as before, using the angle tool for the side you can't get at with the flat one. Note that the angle tool will have to be tilted rather than raise the handle, and this means you must have a really good hold of it, or it will take charge.

Once you have the profiles to shape, proceed with the finishing. This is done with the same tools, but with even finer honing of the edges, much lighter pressure, and at higher speed. Finish with fine emery paper, using finer and finer grades until you are satisfied. If you are going to paint the bulbs, of course, no need to polish. (The originals would be painted, probably, for normal customers but polished for specials!) During this process, slip a tube from the centre of a toilet roll over the column, to keep your fingers out of the holes. Whilst in the lathe, finish off the surrounds to these holes—it is easier done when you can rotate the work than in a vice. Finally, before taking from the machine, check all dimensions.

The studholes in the top and bottom of the casting are marked out from the mating parts—but take care when you break through, both in drilling and tapping. Use paraffin as a lubricant. This job comes later, but I will mention it now. You must take great care that the base, column, and entablature are correctly aligned—you don't want the apertures in the column to look askew. My practice is always to spot through only one hole, drill and tap that, offer all up with studs in place, and if it needs correction, draw the hole with a file. Then spot through with this stud (and a clamp) tightened to hold the parts together. The final job at this stage is to mark out for and file out the the clearances for the crank and eccentric and drill the two little holes in the side which give access to the steam chest bolts. After which, re-prime. I painted my first one so that you could see what it looks like—see fig. 8. Note, by the way, that I hadn't trimmed up the surrounds as I suggested you do, and you can see what happened! (This is the 'Mark I' column, which was bored out at the top; yours will be different, having an external spigot instead.)

Base, Part B fig. 9

Set up in the 4-jaw to machine the underside—you may have to grip by the corners to ensure that it beds down properly. Machine the bottom edge, until the machined face is $\frac{13}{12}$in. from the unmachined part of the top of the casting. Note—when setting up make sure that the lip on the bottom edge is running reasonable true, or it will look odd. Then face the internal machined pad to $2\frac{1}{2}$in. dia. and $\frac{7}{32}$in. below the previously machined edges. Reverse in the chuck and tap hard back to the chuck—use packing if need be—and adjust the position until the four sides are true to centre. Machine the $2\frac{1}{2}$in. dia. face and, if your tool will clear the chuck jaws, the bolt pads as well. Use a sharp knife tool in forming the spigot itself, and undercut the root a trifle. Now take a cut right across, over the cylinder spigot (at present running offset) to give a clean face for marking out.

Set one side of the casting level with a spirit level and scribe a line across at centre-height. Set this line vertical with a square, and again scribe across. Working from these lines you can now mark out for the other holes (except those on the $1\frac{3}{4}$in. dia. face) and for the centre of the cylinder spigot. Figure 10 shows this work in progress.

Now reset the casting true to the cylinder centre, making sure that the casting is hard back on the chuck face or or jaws. The cylinder spigot must be machined parallel to the other faces. Machine and face the cylinder spigot. (If you prefer to machine such spigots to fit the cylinder, the job can be laid on one side and done later, so long as you remember to bed the casting down in the chuck at the time.) Mark out the holes on this face, but leave the two 8 BA ones for the exhaust flange. These are marked out from the pipe itself in due course. Drill all holes. (Leave those for the dowels, of course, until you erect; don't forget them, as I may do in the subsequent description of the assembly!) Re-prime the unmachined parts and set aside.

Plinth, Part E fig. 5, page 18

Run a file over the bottom edge until it sits without shake on a flat surface—machine it if you like, but it isn't essential. File until the overall height is more or less uniform. Set up in the 4-jaw reasonably true to the edges and face the top to dimension. Don't machine the cored hole, but mark out, drill and tap for the 5 BA studs after removing from the machine. These are spotted through from the engine base. If you intend to paint this as a rendered cement plinth, leave the sides taper as cast when filing up, but if you wish to simulate brickwork or ashlar (built up of stone blocks) then you ought to file the sides square to the top. File up the little bevel in either case, and keep the corners fairly sharp. Finally, mark out for and drill the $\frac{1}{4}$in. hole in the side. (This is the entrance to the way out for the exhaust pipe.) Re-prime, and set aside in a safe place.

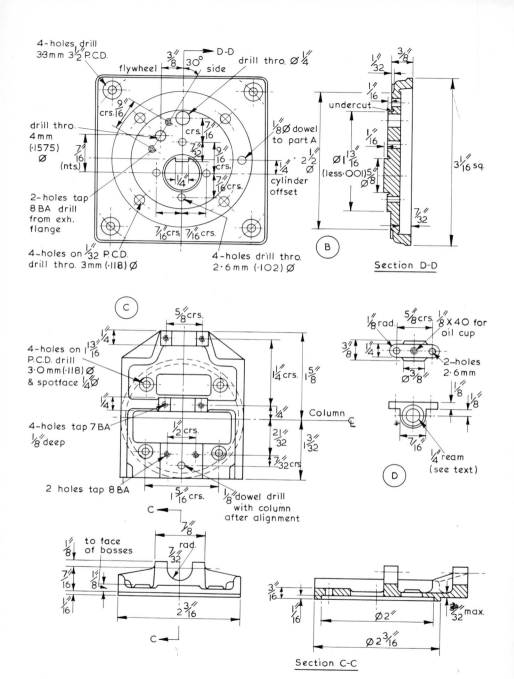

Fig 9-Base and entablature.

Fig 10-Marking out the base.
(The rule is held upright by a small magnet behind.)

Entablature, Part C, fig. 9

The design called for an unmachined area on the underside of this casting, but for production reasons (to keep the cost of your castings down!) the machining allowance will be left on all over, so you must cut the surplus away when you face this side. You are going to need to Araldite the sideplates to this casting, so take care to avoid oil getting on the job if you can.

Set up in the 4-jaw, base outwards. One jaw may have to be reversed for this. Set true to the sides of the casting, and also make sure that the casting lies flat relative to the chuck body. Pay careful attention to the drawing when setting up, as the $\frac{1}{4}$ in. dimension to the main bearing centre is fairly important. This is a fiddly job, as packing is little help, but take some care over it. Machine right across until the thickness of the casting to the unmachined part of the top face is $\frac{5}{32}$ in., and then take off another $\frac{1}{16}$ in.

from the outside in until there is a pad $2\frac{3}{16}$ in. dia. which will be $\frac{5}{32}$ in. thick. You now have to form a socket for the top of the column in this pad. Rough out this cavity undersize, going in about 60 thou., using your feedscrew index to measure with. Now grind your boring tool to a point to leave a sharp internal corner, and finish both the bore and the face of the recess. Take care, as this face and bore are essential to correct alignment. The column should be a snug but not tight fit. Using the same procedure as for the base, set the job square and mark out the centre-lines and then the sides and front— $1\frac{3}{32}$ in. from the centre-line in each case; take care that the sides are parallel, and square to the front edge.

Remove from the chuck, and either file or mill the sides to dimension. Figure 11 shows how I did it, though you are unlikely to have a cutter like that—70 years old, probably! Mill the front edge or face first and this can then be a reference to which the others can be set. The fourth

Fig. 11-Machining the front face of the Entablature.
(The cutter is really a piece of 'Industrial Archaeology'!)

side—that where the outer main bearing housing is—shouldn't need any metal off, but if it does it can be cut off as a final operation. Round off the edge of this part so that it looks 'cast'. Whilst on the vertical-slide you can mark out if you like, using the vertical and cross-slide indexes to get your dimensions, but it's just as easy to do it on a surface plate in this case.

Reset the vertical-slide to face the headstock and set up the casting either in a machine vice or with clamps. (You can't use a bolt in the middle.) Set true to the centre-line with a square from the lathe bed. Machine off the tops of the main bearing housings to correct dimension. If you have a small enough cutter you can also get down to machine the bolt-pads and then do the face on which the governor bracket sits. If you like you can file the lot, and use a chisel on the recessed parts—the firm who made the original may have had a planer, but it's pretty certain they had no milling machine at that date. There is *no reason at all* why

you shouldn't use a file wherever I say 'mill' (or set the thing up in a 4-jaw and face it) and use a chisel where a file won't go.

Once the bearing tops are faced you can mark out from the centre-line for the width of the bearing chops. These can be milled to width, but only $\frac{1}{8}$in. deep; the semi-circular part will have to be filed. Make the width so that the brass extrusion just won't go in, and file the latter to suit—having taken the casting from the machine first, of course. Check that the odd-shaped holes in the base which pass the crank and the eccentric are to dimension, and file them out if not. Mark out for and drill the four no. 32 holes, spotfacing them if you didn't mill them. From these you can spot through to the column (bearing in mind my previous remarks on this process; look at the drawing carefully to see which is front and which back in relation to the cylinder spigot on the base!) and drill and tap that; one more little job done.

Now for the main bearings, part D. Saw these off roughly to length and file the ends square and flat, a little oversize. Mark out for and drill the fixing holes, using a square across the brass extrusion to get them in line. Drill these no. 37, and spot through onto the tops of the main bearing housings in the entablature. Mark each bearing to its correct chop so that they go right way round every time. Drill and tap 7 BA and then fit both blank bearings with temporary screws—steel ones, screwed up tight. Set up the vertical-slide parallel to the lathe bed and attach the casting with clamps and set true to the centre-line. Adjust the vertical and cross-slides until the bearing aligns with the lathe centre—that is, with the vertical-slide face $\frac{1}{2}$in. behind the lathe centre, and the centre-line of the casting at centre height. Fit a Slocomb in the 3-jaw and lightly touch the end of the bearing brass; examine the pop-mark for correct position and adjust the slides if need be. Then centre properly and drill say $\frac{3}{16}$in. through both bearings. Use a straight flute drill if you have one, otherwise reduce the cutting rake on a twist drill. Feed steadily to avoid the drill wandering—don't apply great pressure; just let the drill cut naturally. Now enlarge, to your $\frac{1}{4}$in. reaming size (I use letter D) a step at a time so that each drill has only a few thou. to take out. Finally fit a $\frac{1}{4}$in. reamer and, using the rack hand-wheel, put the reamer through at about 300 r.p.m. Remove the vertical-slide (and the job, don't forget!) and make a little stub mandrel to fit the brasses. Machine the top flanges equally each side of the bolt holes and face the ends to dimension. Form a little countersink each end of the hole. Finally, shape the ends of the flanges as shown on the drawing and refit to the entablature so that they don't get lost.

Preparation for the sideplates comes later, as do the governor fixing holes, but you must now store this part so that (a) it doesn't rust and (b) doesn't get oily. A tin box with a bit of VPI paper in it is best.

Chapter Three

CYLINDER SET

Cylinder, Part J fig. 12

This is a gunmetal casting. Check it with your square and file the top (circular flange, until it lies square to the general body of the casting. Lay your rule on it and decide how much must come off this flange to equalise the thickness of both when finished. (It should be $\frac{3}{32}$ in. thick when the face is $\frac{1}{2}$ in. from the bottom of the port-face.) Set up in the 4-jaw, base outwards, true to the outside of the casting—it doesn't matter if the bore runs out a little. Face the end, using the dimension previously decided, but leave on a few thou.

Use as stiff a boring bar as you can—you may have to change to a larger one if the cored hole won't take a big enough bar—and bore to size (fig. 13). If you haven't a gauge (I use a $\frac{5}{8}$ in. roller from a large roller bearing) use a mike over your inside calipers; this is quite accurate once you develop the feel for it. But pay more attention to finish than to size if you have a choice. For the final cut, hone the tool with an Arkansas stone and with a small radius on the nose; use a cut of not less than .002 in. and auto feed—don't try to take out the odd half-thou. as the tool will soon wear and then rub. If you are troubled with chatter, park a bit of Plasticine on the boring bar, but don't let this rub on the machined bore. Finally, again with a honed tool, finish face the end.

Whilst still in the chuck, mark out for centre-lines and holes other than those for the studs, and carry the mark of the main centres round the outside of the casting to ease later settings. Many model makers prefer to do this job on a surface plate, but the chuck makes a very convenient marking-out fixture. Remove from the chuck.

Make a stub mandrel to fit the bore, set up the cylinder, and face the other end to bring the casting to length—leave it a little over rather than under if you *can't* get such things right! (Rule measurement is good enough.) Whilst on the mandrel, pick up the centre-lines previously marked and scribe these, not too deeply, on this face also. Take off the mandrel and mark out for and drill the holes forming the exhaust passage. Spot-face at the $\frac{5}{8}$ in figure shown. Note that the $\frac{7}{16}$ in. distance should be worked to carefully to avoid breaking into the bore.

The Ports

Now for the portface and ports. Set up the vertical-slide and check it for squareness. Mount the cylinder, top outwards, as in fig. 14 and fit the sharpest milling cutter you have in the 3-jaw. Set the cylinder square to the marks, and take care that you have a good-sized washer under the nut. Engage the leadscrew handwheel to apply cut, and machine the whole portface. You are going to have to 'flat it off' to get a good face for the valve later, I know, but nevertheless, aim for a good finish. With the same set-up you can drill for the ports, using the co-ordinate method and thus avoiding all marking out. For convenience I show the dimensions in fig. 15 redrawn from the top face of the cylinder as reference plane.

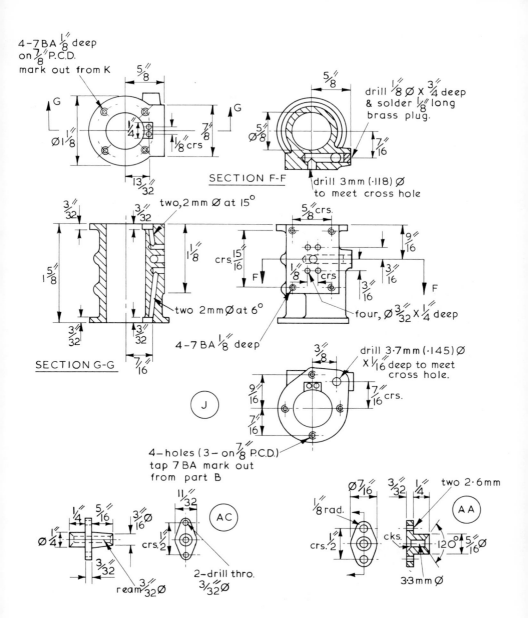

4-7 BA $\frac{1}{8}$ deep on $\frac{7}{8}$" P.C.D. mark out from K

$\frac{5}{8}$"

G

G

Ø1$\frac{1}{8}$"

$\frac{1}{4}$"

$\frac{7}{8}$"

$\frac{1}{8}$ crs

$\frac{13}{32}$"

$\frac{5}{8}$"

drill $\frac{1}{8}$ Ø X $\frac{3}{4}$ deep & solder $\frac{1}{8}$" long brass plug.

Ø$\frac{5}{8}$"

$\frac{7}{16}$"

SECTION F-F

drill 3mm (·118) Ø to meet cross hole

$\frac{3}{32}$" $\frac{3}{32}$"

two, 2 mm Ø at 15°

$1\frac{1}{8}$"

$1\frac{5}{8}$"

crs.$\frac{15}{16}$"

F

$\frac{1}{8}$"   crs

F

$\frac{3}{32}$" $\frac{3}{32}$"

two 2mm Ø at 6°

$\frac{7}{16}$"

SECTION G-G

4-7 BA $\frac{1}{8}$ deep

$\frac{5}{8}$ crs.

$\frac{9}{16}$"

$\frac{3}{16}$"

$\frac{3}{16}$"

F

four, Ø $\frac{3}{32}$ X $\frac{1}{4}$ deep

$\frac{3}{8}$"

drill 3·7mm (·145) Ø X $\frac{1}{16}$ deep to meet cross hole.

$\frac{9}{16}$"

$\frac{7}{16}$"

$\frac{7}{16}$ crs.

J

4-holes (3- on $\frac{7}{8}$" P.C.D.) tap 7 BA mark out from part B

$\frac{1}{4}$" $\frac{5}{16}$"

$\frac{3}{16}$ Ø

$\frac{11}{32}$"

AC

Ø$\frac{1}{4}$"

crs.$\frac{1}{2}$"

$\frac{3}{32}$"

ream $\frac{3}{32}$ Ø

2-drill thro. $\frac{3}{32}$ Ø

Ø$\frac{7}{16}$"

$\frac{3}{32}$" $\frac{1}{4}$"

two 2·6mm

$\frac{1}{8}$ rad.

crs.$\frac{1}{2}$"

cks.

AA

$\frac{5}{16}$ Ø

120°

3·3mm Ø

Fig 12-The clyinder and glands.

Fig 13-Boring the Cylinder.

Fig 14-Milling the Portface.

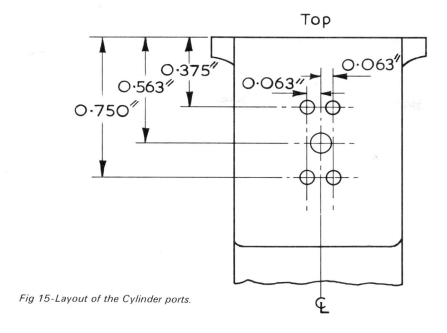

Top

0·375″

0·563″

0·750″

0·063″

0·063″

C L

Fig 15-Layout of the Cylinder ports.

Fig 16-Drilling steam ports with angle-vice.

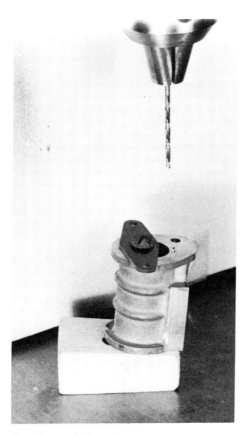

Fig 17-Port drilling using an angled wood block.

The procedure is now well-known, but I will remind you to come up to the final setting on each handwheel in the same direction each time. I always mark an arrow with a felt pen as a reminder! I recommend that you centre all holes with the Slocomb first, and then return to settings to drill—it saves time and reduces the chance of error. Don't centre too deeply, or you will have a port with bevelled edge. If this does happen, just machine a few thou. off the port-face.

Remove from the machine, and set up to drill the steam passages. Figures 16 and 17 show two ways of doing this—and I can assure you that the poor man's way in fig. 17 is just as effective as the other.

Mark out the positions, $\frac{3}{8}$in from the cylinder centre and $\frac{1}{8}$in. apart, and then drill no. 45 vertically for about $\frac{1}{16}$in. deep. Then set up either on the wooden wedge, (fig. 17) in an angle vice, or an adjustable angle plate as shown. I have quite a number of wooden wedges like this, all different angles, to which jobs can be attached quite simply with woodscrews, and very handy they are. The drill will start in the small pilot hole with no difficulty, but they are deep holes for this size and you must clear chips frequently. Take care when breaking through. This having been done you can cut out the recesses in the flanges—mill, chisel, or even just file a bevel. The steam speed is so low (the engine turns on a whiff) that so long as there *is* a passage you won't notice the difference.

Remove all burrs and set the cylinder up on the base, part B. Take care that the cylinder is aligned to the centre-lines and facing the right way round (portface facing the flywheel side). Clamp up, and recheck alignment. Spot through one hole, drill and tap this, fit a screw, and check the alignment again. Draw the hole until it's right (if it's wrong, that is!), fit the screw *and* the clamp, and spot through for the other holes. Drill and tap for these. Note that the drill and the tap will break through half-and-half in the outside of the cylinder wall, and in the soft gunmetal there is a chance of this pulling the hole sideways. Best to set the depth stop on the drill if it has one. Leave all other holes in the cylinder until you have the mating parts.

Glands, Parts AA and AC fig. 12, page 27

These need little description. Chuck by the chucking piece and set true to the casting. Face and centre, then drill no. 30 for AA and no. 41 for AC. Make the 120 deg. recess, and I recommend you use a boring tool if you haven't a 120 deg. countersink, not the point of a big drill. Turn the o/d a few thou. under the nominal size so that it will be a slide fit in the stuffing box, and face the underside

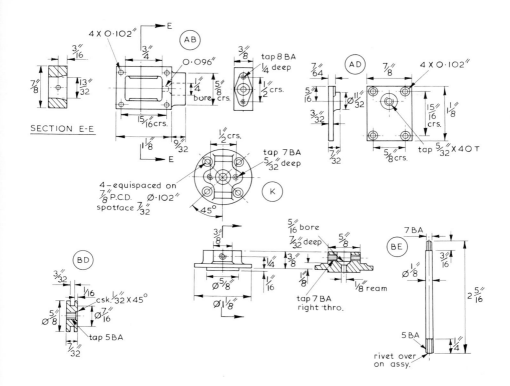

Fig 18-Steam chest, cylinder cover, piston, and rod.

of the flange. Remove and cut off the chucking piece (leave enough on the valve gland to allow for the rod-guide extension) and then grip the other way round in the 3-jaw. Machine the outside, bringing the flange to correct thickness and forming the extension on the valve gland. Countersink the rod-hole. Mark out for the bolt holes and, after removing from the machine, scribe circles for the ends of the oval *before* drilling the holes. On the other hand, drill the holes before filing the oval, so that if one hole runs out you can file the shape to 'look right'!

Steam Chest, Part AB fig. 18

Trim with a file and then grip in the 4-jaw, using packing as necessary, to machine the front and back faces; bring the stuff-

ing box central to the casting when doing this. After machining the second face mark out for the width of the cavity, for the four bolt holes, and for the overall length of the casting. Remove from the chuck and drill the bolt holes, then file or mill the outside. The purpose of the odd shape in the cavity is to guide the valve, so carve it out as shown in fig. 19. The rubbing faces should be parallel to each other and to the long-itudinal centre-line and about 3 thou. clear on the valve. (It is worth trimming the sides of the valve and fitting it to the chest at this stage.) The rest can be chewed out with a coarse file—or your teeth, if you like.

Rechuck in the 4-jaw using lead or other very soft packing to avoid marking the machined faces, with the stuffing box

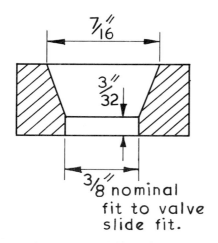

$$\frac{7}{16}{''}$$

$$\frac{3}{32}{''}$$

$$\frac{3}{8}{''} \text{ nominal}$$
fit to valve
slide fit.

Fig. 19-Cross section of steam chest.

pointing outwards. Centre to the sides and faces—never mind if the boss runs out a bit. Face the boss, centre, and drill $\frac{7}{32}$ in. × $\frac{1}{4}$ in. deep. Recentre with a small Slocomb and drill no. 42 (*not* 41 at this stage). Use a tiny boring tool to machine the cavity to fit the gland you have made already; then fit the gland in the hole and run a no. 41 drill right through. This will ensure that the hole is truly central, but don't forget to mark the gland and the boss so that they go on right way round. Remove the chuck and spot through the gland for the stud-holes and drill and tap these 7 BA. You may find the holes in the gland a bit tight—if so, open them out one size. I like close clearance holes but drills don't always behave as they should! Trim round the oval of the stuffing box to make it match the gland.

Chest Cover, Part AD fig. 18, page 31

You must be careful holding this part, it's a bit thin! Chuck in the 4-jaw true to the boss, and set it so that it will machine to uniform thickness. Face the boss and turn its diameter, and take a cut across the bolt-pads. Drill and tap the boss to suit the steam connection you intend to use. The original engine had a length of steam

pipe with a flange on the end, but I think an adaptor for the Stuart 155 combined lubricator and stop valve might be better if you are going to run on steam. Take from the chuck and reverse to face the inner side—don't make this oversize; if anything let the dimension be thinner than shown as there is no room for any excess metal here. File the edges to match the steam-chest.

You can now use the steam-chest as a jig to drill both the stud holes in the cylinder and those in the cover. Use clamps, for two reasons:
(a) To save doing yourself an injury when the drill breaks through.
(b) It is most important that holes match up properly, as otherwise assembly may be difficult.
The piston and rod are dealt with in Chapter V, along with the associated crosshead and running gear.

Cylinder Cover, Part K fig. 18, page 31

As the alignment of the crosshead guides depends on the accuracy of this part, more than usual care must be taken in setting up when machining. The critical points are:
(a) The two flat faces on the gland boss must be parallel to each other and square to the face of the cover flange.
(b) The faces must be equidistant from the cylinder centreline.
(c) The $\frac{5}{8}$ in. dimension must be held to as close limits as possible.

Of these, (a) is the most important; (b) and (c) can be allowed for by adjusting the crosshead slippers if need be. The machining process outlined below will meet these criteria but it is important to check at each stage, always remembering that it is relatively easy to to take off more, but 'putting-on' tools are very expensive.

Set up the cover in the 4-jaw, boss outwards, and rough machine the flange O.D. Face as much of this as you can, and the face of the boss. Centre and drill say $\frac{15}{64}$ in., then bore to a nice slide fit on the gland. Recentre the bottom of the hole with a small Slocomb and drill and ream $\frac{1}{8}$ in.—take the hole right through the chucking piece. Mark out the two centre-lines whilst in the lathe and, though we will not be using them for machining purposes, two lines at the $\frac{5}{8}$ in. dimension of the flat faces. You can also mark out for the four studholes. These are shown at $\frac{7}{8}$ in. PCD but you will find that these lie on the corners of a square of nearly $\frac{5}{8}$ in. side (0.619 in., to be exact!) and it is easier to do it that way.

Remove from the chuck and make a stub mandrel (fig. 20) from a piece of $\frac{3}{4}$ in. or $\frac{7}{8}$ in. material. This is going to serve another purpose later, so let it project between $\frac{3}{4}$ in. and 1 in. from the chuck jaws and make the stub only just long enough to fit the gland. Drill and tap for a 6BA retaining screw if the gland hole isn't a tight fit. (You can always get it off by warming it). Take light cuts and finish machine the O.D. to $1\frac{1}{8}$ in. Face, and form the spigot. NOTE, the face dimension of $\frac{3}{8}$ in. between the face of the boss and the under face of the flange is more important than the thickness of the flange. Remove from the mandrel but leave this in the chuck.

Now turn the mandrel down to make a button as in fig. 20 for use when milling the flats. Make the diameter to 0.625 in. by micrometer and part off. This plug, when fitted into the gland hole, will be a gauge to help get the flats correct. Set up the vertical-slide and take more than usual care to get it square—the working face parallel to the axis of the mandrel, which means you can use a DTI to check with. Fit the gauge you have just made to the cover and clamp to the vertical-slide making sure that you can get at one of the scribed lines to set it truly vertical. I found the set-up a bit of a fiddly job as there wasn't much room to see, but you must persevere and check carefully. It would be wise to have the clamp vertical.

Look at the drawing and you will see a part section showing a shoulder $\frac{1}{4}$ in. below the face of the boss. The width of this is not important, but the $\frac{1}{4}$ in. dimension *must be the same for both faces*—it helps to locate the slide-bar brackets vertically. Proceed as follows. Fit an end mill about $\frac{3}{8}$ in. diameter in the 3-jaw—use paper to get it running true. Check that the clamps and bolts on the

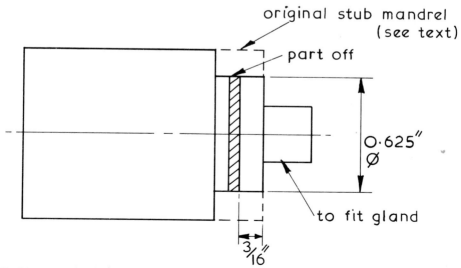

original stub mandrel (see text)

part off

0.625" Ø

to fit gland

$\frac{3}{16}$"

Fig 20-Jig for machining Cylinder cover.

Fig 21-Machining the slide-bar seating on the Cylinder cover.

vertical-slide are tight. Engage the lead-screw handwheel and advance the saddle till the cutter just brushes the $\frac{5}{8}$ in. gauge plug. Note the index reading. Next, operate the cross-slide with the cutter clear of the gauge until the cutter just brushes the face of the gland-boss. Note this index reading. Run the saddle clear, and advance the cross-slide 0.250 in. and lock it. Use the leadscrew handwheel to put on cut (I suggest 0.010 in. at a time) and then traverse with the vertical-slide repeating until you reach the index reading previously noted for the lead-screw handwheel. When you get within a few thou of this, keep an eye on the gauge plug. If this shows a witness that the cutter is touching it, stop there. Your mandrel may expand in use, there may be slack in the leadscrew thrust, or in the half-nut, but the fact that the cutter is touching the plug is all that matters. So, take off the last few cuts a thou or so at a time. (See fig. 21).

Now *without altering the cross-slide*

turn the workpiece round to offer the other face to the cutter. This time, though, square up the work by using your try square against the face you have just machined, ignoring any centre-lines. Take your usual meticulous care over the job! Then repeat the machining procedure on this face as before. The critical faces will now be within a thou or so of the dimension desired, truly square and parallel. One thing, though. When did you last check the squareness of your vertical-slide? Some of them aren't!

Before removing from the machine, set the work with the boss centre at lathe centre height and scribe a line across the faces to mark for the 7BA tapped holes. Remove from the machine and show the faces to your micrometer. If they lie between 0.623 in. and 0.626 in. leave well alone—so long as they are parallel. If not, then make a note and remember to look at it when you come to deal with the upper guide supports and the crosshead slippers. You can now drill for the four studholes, spot through and drill for the gland studs, and drill and tap the two 7BA holes in the faces. These (and the gland studs) must be truly upright; it doesn't matter that the hole will break through into the gland cavity, but clean off the burr. Spotface the nut seatings to $\frac{3}{32}$ in. thick, clean up, and remove all burrs.

Clearly all this accuracy will be wasted if the cylinder cover is not properly aligned on the cylinder, so the positioning of the studs in the latter is important. Lay the cylinder, portface down, on the lathe bed and lightly clamp the cover to it. Align this with a square against one of the crosshead faces and tighten the clamp. Check that nothing shifted. Upend the assembly and spot through no. 37; mark the cylinder and cover so they will go together right way round, and then drill and tap the four holes. And that is that—except that when you come to make the joint for this top cover you must see that there are no wrinkles or bits loose on it, which might throw the cover over. An electric iron will help.

Chapter Four

CRANKSHAFT
AND FLYWHEEL

Crankshaft, Parts CA & CC fig. 22

The shaft part should have centres in the ends when you get it, but if not, set up in the 4-jaw and centre both ends. Then turn up your soft (headstock) centre true, set a test bar between centres, and adjust your tailstock to turn parallel. Don't try to use the top-slide instead, as you must use power traverse to the tool to get a good enough finish on this job. Put the crank between centres and machine to $\frac{1}{2}$ to 1 thou. under reaming size—or to fit the bearings etc. if your reamer is tired and heavy with years. As I said, use power feed at all times, and if you are afraid of running into the headstock, set the tumbler reverse to traverse to the tailstock! Another tip, if your machine also has seen better days and won't turn parallel any more. Chuck a piece of 1 in. b.m.s. in the 3-jaw, say with 2 in. projecting, and turn a 60 deg. centre on it. This will throw the work further up the bed, to a place perhaps less worn. After turning the shaft, machine the back of the web to the recess shown on the drawing. Get a good finish on the $\frac{1}{2}$ in. dia. shoulder. Face off the end of the shaft, leaving the centre in; the exact length is not all that important.

For the next step you can set over in the 4-jaw if you like, to machine the outer side of the web and drill the hole for the crankpin, but you will, in effect, be gripping with only two jaws, and truth cannot be assured in such circumstances. So, find a chunk of metal about 2 in. dia. and say 2 in. long and make up the jig shown in fig. 23. Anything will do so long as it is larger than the hole in your 4-jaw and is true on the outside (parallel, that is). Chuck this piece and face one end. Reverse, and tap this face back against the face of the chuck (which you should clean first) the piece being offset in the jaws by about half the crank throw. Face the exposed side, centre, drill no. 1 and then bore as deep as you can go to be a good fit on letter 'D'—or whatever drill you use for reaming to $\frac{1}{4}$ in. This short bored section acts as a guide to the reaming drill and helps keep it true. Drill and then ream right through. Run at about 200 r.p.m. for mild steel, use lashings of cutting oil, and clear chips frequently. Mark no. 1 jaw-seat on the piece, and then take it out.

Finish the component off as shown in fig. 23. The two little pegs are not esential, but I have shown them to save readers the trouble of writing to say that they should be there. Note the little copper plug under the setscrew which prevents marking the shaft. Return to the chuck, but offset the piece the opposite way so that the hole now runs a full crank throw off centre. See that the block is hard back against the chuck face, and adjust the jaws till a piece of silver steel stuck in the hole runs off by the correct amount; at the same time run a dial indicator along the length of the $\frac{1}{4}$ in. rod to ensure that it is pointing parallel to the lathe bed—both ways. If not, adjust. If you get the offset correct to .005 in. this should do—there is ample clearance in the cylinder. Once satisfied, set the crank in the hole, adjust its position until the centre of the small end of the web is true, and tighten the setscrew. Machine the outer face of the web to dimension. The $\frac{5}{16}$ in. dia. boss will act as a guide for filing the outline later.

Set a piece of packing between the web and the face of the fixture to take the drilling thrust, centre, and drill, first no. 32 then 31, and finally ream. If you haven't a $\frac{1}{8}$ in. reamer, drill to this size after the no. 31. The exact dimension doesn't matter, as you can make the pin to suit. Put a light countersink on this face, then remove from the jig and countersink the back. The web can now be filed to shape. Apart from the machined thrust faces, all edges should be rounded off—the original crank would have been a forging and this is what it should look like in the model. Avoid deep file-marks of course—scale them up × 12 to see what they would be like in real life—but don't aim at a mirror polish. The original crank would have been unmachined and probably painted red.

The crankpin is a simple turning job, but make the $\frac{1}{8}$ in. peg either a good press fit to the hole in the web or an easy fit and use Loctite if you like that method. In either case you will rivet the end of the pin over. Get a good polish on the crank-pin itself, and aim for a sharp corner at the junction with the $\frac{1}{8}$ in. peg, to ensure that it seats flat and true to the crankweb. If you use Loctite, don't forget to degrease both pin and web; if a press fit, don't degrease—rather put a spot of grease on the peg; horses for courses! Finally, rivet over the pin, making a neat job of it; you can leave the rivet head to show or file it off as you please. Reverting to the fixture, I now find that the photo I took is a blank, but fig. 24 shows a similar (smaller) one which will give you the general idea.

Eccentric, Part AK fig. 22 page 39

The Eccentric is an iron casting with a chucking piece. Many ways of dealing with eccentrics have been described—use the one you fancy most. For a change I departed from my usual method of calculated packing in the 3-jaw and did it like this. After trimming, hold by the body and true up the chucking piece. Now grip the latter in the 3-jaw and turn the O.D. to correct diameter less say one thou clearance. Form the groove with a narrow round-nose tool, setting it a little towards the eccentric boss so that it will be central when finally machined all over. Take care to get a really good finish on the eccentric —friction at this point is one of the main losses in a steam engine.

Return to the 4-jaw, again gripping by the chucking piece and set it off centre by the designed throw ($\frac{5}{64}$ in.). Use a dial indicator to get it right, remembering that the DTI will give an indication of TWICE the offset. If you haven't a DTI use the cross-slide index as a micrometer 'touch gauge'. Machine the boss, taking light cuts and face this and the side of the eccentric shave. Centre, drill (stopping when you get to the chucking pice) and bore to size. Take from the chuck, saw off the chucking piece, and then hold by the boss and face the outside. Finally, mark out for and drill for the two 7BA set-screws.

Flywheel, Part H fig. 22 page 38

So far as can be seen from the catalogue the flywheel was not machined, a sep-arate belt pulley being used, so there is very little machining allowance on the rim. But it may need a bit of trimming here and there, and the casting should be so true that if you *must* skim the rim you need take off very little metal. Before machining, do all the preparatory work for painting on the spokes etc., and then clamp (fig. 25) to the faceplate setting the rim dead true. Don't clamp too hard and make judicious use of packing to avoid distortion. Face the boss and then drill and ream (or, better, bore) $\frac{1}{4}$ in. There is no need to face the other side of the boss, but if you do I suggest doing it on a stub

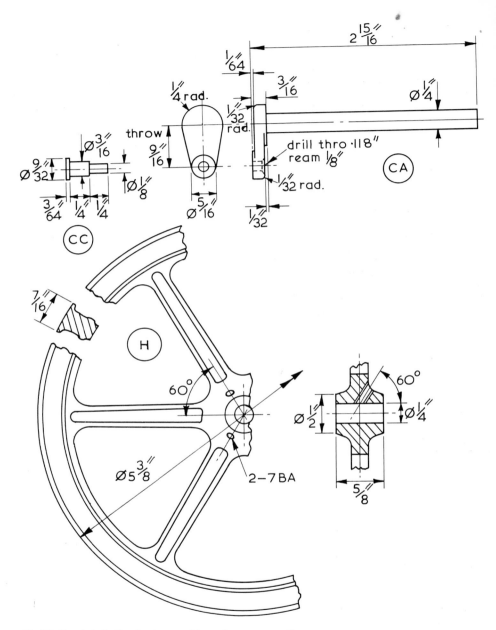

$2\frac{15}{16}''$

$\frac{1}{64}''$

$\frac{3}{16}''$

$\varnothing\frac{1}{4}''$

$\frac{1}{4}$ rad.

$\frac{1}{32}$ rad.

throw

$\frac{9}{16}''$

drill thro ·118″
ream $\frac{1}{8}''$

CA

$\varnothing\frac{3}{16}''$

$\varnothing\frac{9}{32}''$

$\varnothing\frac{1}{8}''$

$\frac{3}{64}''$ $\frac{1}{4}''$ $\frac{1}{4}''$

$\frac{5}{16}''$

$\varnothing\frac{5}{16}''$

$\frac{1}{32}$ rad.

$\frac{1}{32}''$

CC

$\frac{7}{16}''$

H

60°

$\varnothing 5\frac{3}{8}''$

2-7 BA

60°

$\varnothing\frac{1}{2}''$

$\varnothing\frac{1}{4}''$

$\frac{5}{8}''$

Fig 22-Crankshaft, flywheel, eccentric and governor pulley

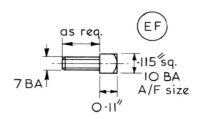

as req.

7 BA

·115″sq.
IO BA
A/F size

O·11″

SETSCREW FOR ECCENTRIC
AND FLYWHEEL 4 b.m.s.

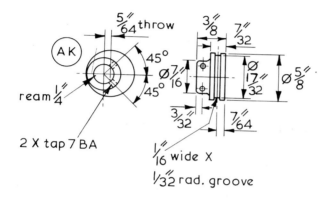

$\frac{5}{64}$ throw

AK

ream $\frac{1}{4}$

2 X tap 7 BA

45°

Ø $\frac{7}{16}$

45°

$\frac{3}{8}$ $\frac{7}{32}$

Ø $\frac{17}{32}$

Ø $\frac{5}{8}$

$\frac{3}{32}$ $\frac{7}{64}$

$\frac{1}{16}$ wide X

$\frac{1}{32}$ rad. groove

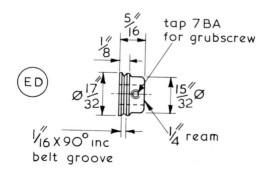

$\frac{5}{16}$

$\frac{1}{8}$

tap 7 BA
for grubscrew

ED

Ø $\frac{17}{32}$

$\frac{15}{32}$ Ø

$\frac{1}{4}$ ream

$\frac{1}{16}$ X 90° inc
belt groove

Fig 22-continued

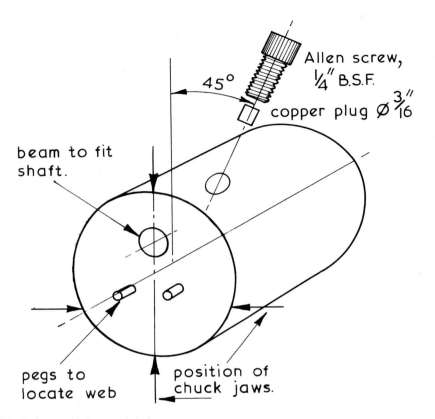

Allen screw, ¼" B.S.F.

copper plug ∅ 3/16"

45°

beam to fit shaft.

pegs to locate web

position of chuck jaws.

Fig 23-Jig for machining crankshaft.

Fig 24-Typical jig for machining single-throw crankshafts.

mandrel. Finally, drill and tap for the set-screws, and re-prime āll the parts you have filed up.

Governor Drive Pulley, Part ED fig. 22

A straight turning job, but there isn't much room where the pulley fits so pay some attention to the OD. The groove must be the narrow vee shown or the cord won't grip, so you will have to grind a tool to 45 deg. included angle. I recommend that you bore the hole a good fit to the crank rather than ream. The grub screws must not project more than half a thread when assembled.

Fig 25-Machining and boring flywheel boss.
(If the rim requires skimming, packing must be set behind.)

Set Screws, Part EF fig. 22

These are made by filing carefully the head of a hexagon head setscrew to fit your 10 BA spanner. Don't forget to bevel the corners, as on a square nut. They must then be shortened—especially those on the eccentric—so that they foul nothing. You may find that you have to use a normal slotted grubscrew in the governor drive pulley, as a headed one may not clear at all.

At this stage, assemble the shaft into the main bearings and entablature. If you find it very stiff to turn, check first that the bearings are the right way round—the same way as when first boring them. Then make sure that both bearings have a slight clearance in the entablature. But better too tight than too slack—you should be able to turn the shaft with a fair pressure of two fingers on the crank-web. If it's very tight, and you can find no fault (assuming the shaft is free in the bearings individually) you may have to put the reamer through again. If only moderately overtight, try running the shaft in the lathe, using plenty of oil. Finally, check that the crank doesn't foul the entablature anywhere. If it does, first make sure you have all to dimension and if that's all in order, file away the offending part of the casing. A clearance of 0.015 in. should suffice.

Chapter Five

RUNNING GEAR

Connecting Rod, Parts BF to BK fig. 26A & B

At first sight this looks a formidable piece of work, but I assure you that it's easier than it looks! However, you will in due course need a piercing saw—a metal cutting fretsaw—and a fine Swiss file. I recommend a no. 1 or no. 3 saw blade; the no. 5 is much stronger, but needs a larger hole to start with. The best file is an 'Escapement Warding', which is thinner than the usual Swiss file. You must expect to break a couple of saw blades if you are not used to them.

The stock provided is over-length, as you need a bit in which to drill a hole for the 'big end'—you can't drill half a hole—and something the other end to hold it with. It is also over width, but the correct thickness. Clean up the exterior with fine emery and leave a matt surface to accept scribing lines and then set up in the 4-jaw, as true as you can manage, and centre both ends. It is important to get these centres in the middle of the $\frac{3}{16}$ in. width. Now mark the approximate positions of the holes and the ends of the taper part with a felt pen, bearing in mind what I said about the extra length at the big end. Set up between centres, large end at the tailstock and rough down all the rod except that needed at the big end (say a length of $\frac{7}{8}$ in.) to just over $\frac{5}{16}$ in. dia., which is the approximate dimension across corners at the crosshead end. Then take a roughing cut down to 0.2 in. dia. over the length which will be the taper shank, leaving say $\frac{3}{32}$ in. short at each end.

Check the centre-pressure, as the job may get warm. Now mark out exactly the position of the ends of the taper shank.

Find a suitable round-nose tool and hone the end to a very fine polish. Set the top-slide over to the required angle—a shade under 1°—and lock the saddle. Machine the taper using the top-slide until the large end is just $\frac{3}{16}$ in. dia.—let the other end look after itself. Once the taper is done, reset the top-slide and take a finishing cut over the small end of the workpiece to reduce it to 0.308 in. dia. This is the exact across corners dimension, and the edge will guide you when filing it to shape later.

Slightly increase the centre-pressure and set the flat side of the rod vertical with your square, so that you can mark out the longitudinal centre-line—do this both sides. Remove from the lathe, and mark the exact position of the large end hole from the end of the taper. From this, strike a radius of $2\frac{7}{8}$ in. to get the centre of the crosshead pin hole. Make small centre-pops, examine these under a glass to see that they are in position; deepen them with a small drill and re-examine, correcting with the punch if needed. If your drilling machine is true you can drill both holes in this, otherwise clamp to the lathe faceplate and drill from the tailstock. In both cases start with a small drill and enlarge slowly. The $\frac{5}{32}$ in. hole is reamed to fit a bush, the other end can be drill size. Return to the lathe between centres, and turn the large end down to 0.338 in., but don't forget you have very little metal round the hole! Take it easy.

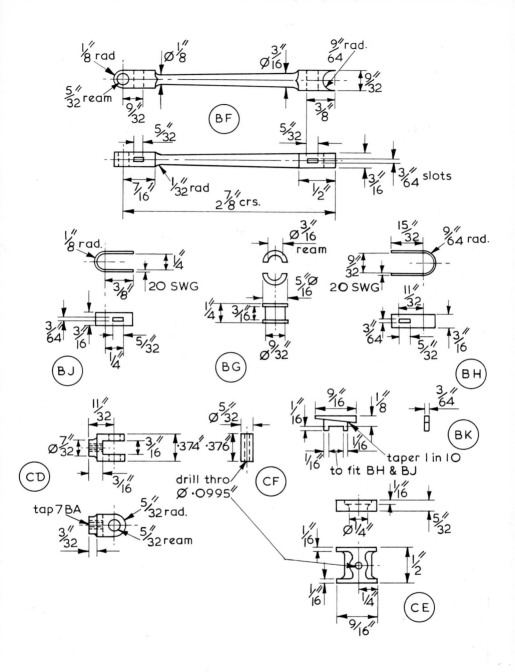

Fig 26A-Details of connecting rod and crosshead.

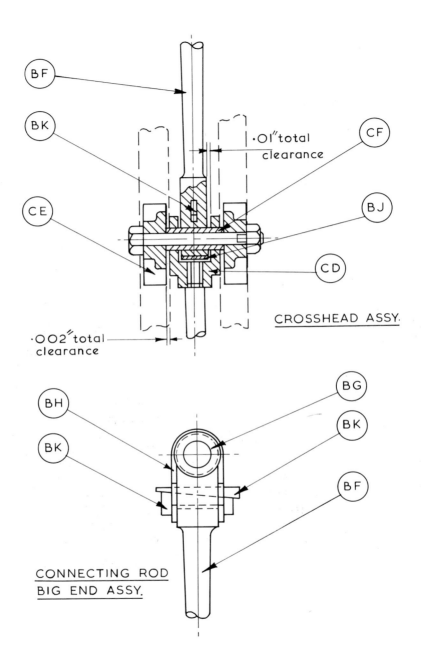

BF

BK

CE

·OI″ total clearance

CF

BJ

CD

CROSSHEAD ASSY.

·OO2″ total clearance

BH

BK

BG

BK

BF

CONNECTING ROD
BIG END ASSY.

Fig 26B -Assembly of crosshead and connecting rod.

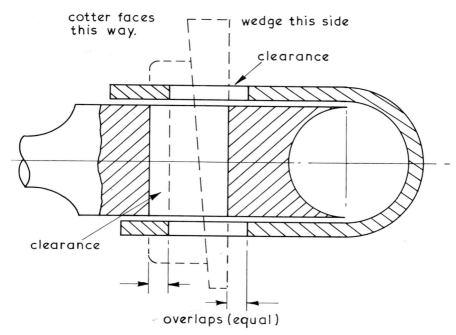

cotter faces this way.

wedge this side

clearance

clearance

overlaps (equal)

Fig 27-Connecting rod bearing, showing cotter clearances.

Remove from the machine and saw off the surplus from the 'large' end leaving a bit to file down to make the 'half-hole' exactly half. File to width, and then take off just a shade from the edges of the half-hole so that there is a tiny flat there. File the crosshead end to width and form the radius on the end. This must not be oversize—there isn't room in the crosshead for any extra metal. Finally, reduce the width of the small end to 0.175 in. to give the necessary side clearance in the fork.

Now for the cotter holes. (See fig. 27). Mark the position carefully, and centre-pop for three holes, one at each end of the slot, one in the middle. Drill the end ones first, no. 57, then the one in the middle, but don't worry if this one runs awry—it's only to save sawing time and *can* be left out. Put the blade in your piercing saw, cutting edge facing the handle, and thread it through the hole before attaching the other end and putting on plenty of tension. Carefully saw down, first one side of the slot, then the other. Take your

time, and use plenty of lubricant—3 in 1 or similar—and little pressure; let the saw cut on its own. The blade is a carbon steel cutting tool and has a low speed, so don't stroke like a pneumatic drill. Take it easy! Once you have sawn out, finish the slot with the little file. A piece of 18 gauge stuff should slide through but not wobble. Holding the job may be a problem, and I use a hand or instrument vice held in my normal bench vice for this class of work.

Big-End Bearings, Parts BG, BH, BJ fig. 26 page 43

The split bush, BG, calls for little comment. I hope you will get a piece of half-round stuff to solder together, but if not, saw the stock in two, clean up, and solder together. If you have no $\frac{3}{16}$ in. reamer drill one size down and finish with a $\frac{3}{16}$ in. drill; form a little bevel at the ends of the holes. The groove in the outside is turned to

Fig 28-Fitting cotters, using a small instrument vice.

fit the half-hole in the rod. When parting off, leave a shade on so that the width can be filed to fit the crankpin. Don't unsolder until you are ready, but I suggest you mark the mating parts (and the rod) so that it goes on the right way later.

You can now make the two straps, parts BH and BJ. In doing so, the slots must be positioned as shown in fig. 27 to ensure that the taper cotter tightens up on the gib. (That at the small end is a dummy, of course.) The material provided is pretty soft and bends easily. Mark out for width and trim to this, removing burrs, and then cut off pieces a shade over length—the longer the better, in fact. Fold over a rod of the appropriate size, and take the bend just a little too far—it should embrace the rod, not just touch it! Try and get the bend in one smooth action to avoid kinks, but if you do get a 'not round' bend, true it with a plastic mallet on the back. Mark out for the slots (see fig. 27 again) and drill as before, with packing in between the legs of the 'U'. Saw and file them, but keep trying them

on the rod (with the bearing in place at the large end) and ensure that you can just see a little of the rod visible at the end of the slot at the 'U' end of the strap, but overlapping the rod at the other end.

The cotters BK are a simple filing job but the gap between the shoulders of the gibs should be a good fit on the straps when in place—should need pushing on. Note that the large end of BOTH cotters should be on the right when looking at the front of the engine. Make the wedges overlong, and cut off when the job is finished. At initial erection only the very end of the thin end of the wedge should show, (see fig. 28) but will have to be knocked in a bit after bedding down. When fitting the split bush to the crankpin and rod, adjust the clearance until it just feels tight. If it swings about, it's too slack, but if it won't fall down under its own weight, about 45 deg., too tight. Don't forget to check alignment, by the way. Fit a rod into each bearing (no bush as yet in the small end) and see that they lie parallel both ways. (See fig. 29). If they don't, twist or bend the rod until they are. (Don't use the test rods to apply leverage though!) When finally fitting the cotters, by the way, the large ends of the wedges should be on the same side at each end, and on the right when looking at the front of the engine.

Piston and Rod, Parts BD and BE
fig. 18 page 31

Chuck the piston by the chucking piece with sufficient projecting to ensure that you can get a parting tool in. Face, turn the O.D. to say .005 in. oversize, and rough out the groove. Centre and drill no. 38, right through, clearing chippings frequently to ensure that the drill doesn't wander. Tap 5BA guiding the tap from the tailstock chuck. Go in a few threads with the taper tap, follow with the plug and finish say 5 threads, then repeat with the taper for 5 threads and so on. This way there is less risk of the thread going off true. Take a final skim off the face, and finish turn the groove; then part off almost

to correct thickness—leave a trifle on for finishing when fitted to the rod. To make the latter, part off to $\frac{1}{32}$ in. over the correct length and chuck in the 3-jaw. Check for truth and fit paper if it is more than a thou out. Turn down one end to 7BA—make it .095 in. to allow for the way the metal squeezes up into the die—and screw with the tailstock dieholder. Reverse in the chuck, and screw the other end $10\frac{1}{2}$ threads × 5BA. Degrease, anoint with Loctite, and screw on the piston. Get it hard on, and then lightly rivet the end of the rod. Don't belt it—just a lot of gentle taps till you have a nice overlap of the hole. Trim the O.D. down to a slide fit in the cylinder, and face the end. (This will be the parted-off face) taking off the rivet head too if you like.

Now, if your chuck runs out so much that paper doesn't help, or, worse, if it holds the rod on the slant, make a little blind collet. Chuck a piece of $\frac{1}{4}$ in. brass or aluminium, about $\frac{1}{2}$ in. long, grip only enough to hold it firm, and face the end. Centre, drill, and ream this very carefully to fit the rod. Put the rod in place, and

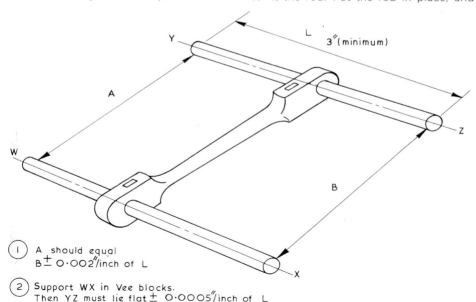

① A should equal
  B ± 0·002"/inch of L

② Support WX in Vee blocks.
  Then YZ must lie flat ± 0·0005"/inch of L

Fig 29-Mandrel and trammel test on connecting rod.

add a little pressure on the chuck jaws; this should grip the rod firmly enough for the final machining of the piston diameter, and will be true.

Crosshead, Forked End, Part CD fig. 26 page 43

Chuck the piece truly in the 4-jaw, face the end and turn down the $\frac{7}{32}$ in. boss, a full $\frac{3}{32}$ in. long. More rather than less. Whilst set up, mark out the centre-lines on sides and end. Set up the vertical-slide and machine vice, slide facing the headstock, and grip the work in the vice, setting the sides truly square to the face of the slide. (The faced boss should be facing the headstock.) Adjust the cross- and vertical-slides until the boss centre is at centre-height, centre with a Slocomb, drill no. 46 to $\frac{7}{32}$ in. deep and counterbore no. 39 about 40 thou. deep. Don't move the vertical-slide. Tap the hole 7 BA, using the 3-jaw to guide the tap truly. Release the work in the vice and turn it through 90 deg.—a screw in the tapped hole will help in manipulation. Make sure that it is at right angles. Set the Slocomb in the 3-jaw and adjust the cross-slide until it makes a mark $\frac{11}{32}$ in. from the faced and tapped end. *Don't* move the vertical slide. Centre, drill, and

Fig 30-Using guide buttons to form forked end.

ream $\frac{5}{32}$ in.—do the best you can with a $\frac{5}{32}$ in. drill if you haven't a reamer, but go under rather than oversize.

If you have a slot drill the right size you can mill the slot in this set-up; simply turn the piece round through a right angle again and after checking settings, cut out say 10 thou. at a time until you reach the right depth. You will be using the vertical-slide hand-wheel to put on cut. Otherwise you can file it. After slotting, remove all burrs, lightly countersink the cross-holes both ends, and file the profile to a neat semicircle, using filing buttons as seen in fig. 30. These are just discs of silver steel, turned to the correct O.D., drilled to suit the hole in the work, hardened, and assembled with the work-piece with a bolt. They must be free to turn but not wobble, and you just keep on filing until the file runs on the buttons.

The final job is to reduce the width of the fork to one thou. under $\frac{3}{8}$ in., but before doing so, check the right-angularity of the holes and the slot. If it's a lot out, I'm sorry, but you will have to start again, but if it's only say .005 in. in a 3 in. test bar, not to worry. If the holes are square but the slot not, this can be corrected with a file on erection.

Crosshead Slippers, Part CE fig. 26A page 43

Set true in the 4-jaw, chucking piece outwards, and machine both the chucking piece and the exposed face of the casting. There are two of these, so do both. Change to the 3-jaw, and hold the slipper by the chucking piece; face to the correct thickness, centre, and drill no. 39. Then very lightly countersink the hole. Remove and saw off the chucking piece, after which the boltface should be spotfaced or filed. For the next operation you need a setting gauge again. Simply turn up a disc about $\frac{3}{16}$ in. thick and 0.503 in. dia.—a 'shade over half inch', that is—and drill no. 39. It is important that it be parallel, so do it by turning, facing, and parting off. Bolt the two slippers together, 'outside out' with this washer in between, and set up in the vice on the vertical-slide. Mill the first pair of faces until the cutter just brushes the washer, then reverse (making sure that the machined face is hard against packing on the vice back) and mill the other. You need to take care that the two faces are parallel, but the washer ensures they are the correct dimension. You can use the same technique if you are filing them. Finish by filing the profile to shape. Whether you paint them or leave them as bright metal is up to you. I painted mine red.

Bush and Bolt, Parts CF and CG fig. 26A page 43

This bush must be a working fit in the connecting rod and a 'not tight' fit in the forked end, which may take a bit of juggling with dimensions. Study the assembly sketch shown on the drawing and at fig. 26A and you will soon get the idea. It doesn't matter if it is a working fit in both, but it mustn't be slack in either and definitely not a tight fit in the fork or you will never get the engine to pieces again. Note that the length must be 0.002 in. greater than the width of the fork—make it a bit longer and file down to size, using your micrometer. The bolt will come ready made, but may be a bit tight in the hole. If it is, just ease it with a file. Make a check assembly of the fork, slippers, bush and bolt and mark adjacent parts so that they always go on the same way.

Chapter Six

VALVE GEAR

Eccentric Strap, Part BA fig. 31

This engine ran very slowly in the original, and looks better running slowly as a model (though it *will* take off to 1500 r.p.m. if you let it!) For this reason I have designed it with small passages, a somewhat restricted 'way out' for the steam, and a pretty late cutoff with zero lead. Set properly, these valve events give very steady slow running and a nice even and sharp beat. She uses so little steam that there is no point in seeking economy with an earlier cutoff. However, at this setting it IS fairly important to avoid any slack in the gear. The valve-nut must fit sweetly and if there is any slackness in the eccentric rod pivot bolt, make a new one to fit the hole properly, if necessary sizing the hole in the rod and in part AF (fig. 18) to suit.

The eccentric strap comes as a single casting, and the first job is to divide it into two, keeping an eye on the drawing so that the cut is in the right place. File or mill the faces smooth and solder together with soft solder. Get a good joint as by the time you have finished there will be precious little 'joint' left! Set up in the 4-jaw, so that the joint runs at centre-height when horizontal, and the outside of the casting is true the other way. Face the outer side. Bore the hole till it is a close slide fit on the eccentric sheave, Part AK fig. 22. Whilst still in the chuck mark out for the height of the 'ears' and the vertical centre-line. You can also scribe a line round at $\frac{13}{32}$ in. radius to give a guide when filing the profile later.

Take from the chuck, and make a stub mandrel to fit the strap; not a tight fit, or the solder will give way. If it slips, rig up a clamp on one of the chuck jaws to act as a driver. Machine to width and then scribe a vertical centre-line on this side also. Remove from the mandrel, but leave the latter in the chuck. File the ears to the correct height and the boss to which the eccentric rod will be attached—this last to be flat as well as to size. Return to the mandrel, set the vertical centre-line horizontal, and mark out for the two 10BA screw holes. Mark the two halves so that they go back the same way. You have finished with the mandrel now so remove it.

If you have a good drilling machine and vice, then you can simply mark out crossways for the bolts, and then drill and tap; drill clearing size until you reach the join and then follow tapping size, and tap with the top end acting as a guide. If your machine is not so hot then it will pay to set up the vertical-slide with a vice to hold the workpiece, and drill from the lathe chuck. You can drill and tap for the eccentric rod at the same time too. Finally, file all round to profile—fit the 10BA screws, or it will come apart—making sure that the overall width is correct or under-size; not over or it will foul the entablature. $\frac{15}{16}$ in. overall is correct.

Eccentric Rod, Part BB fig. 31

I think you will find this comes with the eye end ready drilled, as I have used a standard part from another engine. It

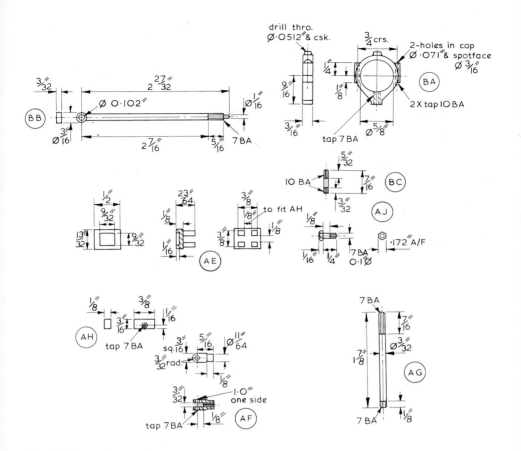

Fig 31 - Valve and operating gear.

may even come finished, in which case it may need altering as to length! This may seem an odd procedure, but with wages at their present level it is far cheaper to use an existing part made on automatic machines than to have a chap cutting up odd bits of rod to send to you for you to make it.

Assuming that the eccentric rod is NOT finished, chuck the rod with about $\frac{5}{16}$ in. projecting and screw it 7 BA with the tailstock dieholder. It should be a tight thread in the eccentric strap you have just made. Remove and offer to this strap, and screw it in until the centre distance between strap and eye end is correct. Cut off the surplus sticking into the bore of the strap, but leave about $\frac{1}{16}$ in. on for the little pip. Return to the chuck and machine this pip to an easy fit in the groove in the cast iron sheave. You may have to screw the rod a bit more to accept the locknut at this stage. Drill the eye end, undersize, and open up one size at a time until it is a snug fit on the pin part AJ. Then file the width of the eye to fit the forked end of the valve rod. Finally, reassemble to the strap and set the centre distance and lock with the locknut.

If the rod comes finished, grip it by the shank in the 3-jaw and part off any surplus length. Rethread if need be (using your tailstock die-holder) and then turn down the little pip at the end.

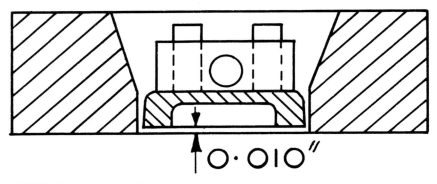

Fig 32- 'Lift' of valve from Portface.

Measure the hole in the eye—use a drill shank as a gauge—and adjust the diameter of the pin AJ and the plain hole in the fork AF to correspond. You may have to open out BOTH holes to get them exactly the same.

Valve and Rod, Parts AE to AJ
fig. 31 page 51

The valve comes as a hot pressing, not a casting, and you will find it nearly spot-on to size. Rub the valve face on a fine flat file and then either with a little scraper or fine emery paper bring it dead flat, checking with marking blue. Fit the sides to the valve-chest making sure the sides keep square to the ends. The latter need no more than a polish to get a sharp edge, and if it is then a shade over length, don't worry. I have designed with a very late cutoff so that the engine can run really slowly when on exhibition, and a bit of extra on the valve length will do no harm. You must now prepare the valve-nut, part AH, and fit to the slot in the valve. You will get a little bit of material for this which fits as it is, but I always find it difficult to tap such a small piece dead square. So, chuck a piece of round brass bar, face the end, centre, drill, and tap, guiding the tap itself from the tailstock chuck. Part off a few thou. over thickness and then file it to a more or less rectangular shape—the rounded ends don't matter.

Fit this to the slot in the valve; first clean the faces of the slot until it is parallel, then file the parted-off face of the nut until it is a nice slide fit. You may have to deepen the slot a bit, but it is just as easy to file a bit off the nut—the object of this is to ensure that when assembled the valve can lift say ten thou. off the port face to let out any water that may be trapped in the cylinder. I suggest you leave this operation until you have made the valve rod, then you can try the valve, nut and rod in the steam chest. For the valve rod, first part off to dead length and then screw each end in turn using your tailstock die-holder. Make the short thread which goes into the forked end a trifle large so that it is a tight fit.

For the fork Part AF, you are supplied with a bit of $\frac{3}{16}$ in. square stock. Grip in the 4-jaw and set true, then face the end, turn the little shoulder, drill, and tap—again guiding the tap from the tailstock chuck. Mark out and cross-drill tapping size for the pin hole, then open up halfway through no. 42. Tap from that side, so that the 42 hole acts as a guide. The slot is simply formed with a warding file, but you can use a slitting saw the right width if you are prepared to take a long time over it. Finally, round off the end to look nice. NOTE, at the time of writing it is not clear whether the bolt, part AJ, will come ready made, or a piece of stuff to make it from. The latter operation is a kid's practice job, so I won't describe it, but in either case, check the diameter of the parallel part and if need be drill the fork, and the eccentric rod, to suit the bolt diameter with no slack.

Now try both valve and rod in the steamchest with the gland in place—first putting a reamer through with the gland in place on the chest. If the threads bind in the gland, file them down with a fine file at the *valve end*. Your die has probably squeezed the metal up a bit. Check that the valve has a few thou. clearance at the sides—0.003" a side is ample. Finally, lay the valve and chest on a flat surface, as already mentioned, and check that the valve can lift. (See fig. 32). Finally, mark the valve, nut and steamchest so that they are always assembled the same way round.

I'm afraid you will have to make the 10BA studs for the eccentric strap (Part BC) from the piece of material provided. Use your tailstock dieholder and do the job properly, not forgetting a turned finish on the exposed end. *Don't* be tempted to use that little bit of 10BA screwed rod you have handy! Not done in the best of circles, though some authorities assert the contrary!

Chapter Seven

PRELIMINARY FITTING AND ERECTION

Yes, I know the governor is yet to make, but this will not affect the running of the engine, so there is no reason at all why you should not 'see how she runs'! In any case it is always as well to erect as soon as the engine can be turned over, (a) to see that all is well and (b) to make sure you haven't any bits yet to make. Which, as it happens, there are, so let's deal with them now!

Slide Bars, Parts CH, CJ, CK
fig. 33

Prepare the bars first. You will have to reduce the thickness, I'm afraid, as material of the right size is, for the time being, unobtainable, though it is hoped this may be available by the time this is in print. In reducing thickness, I recommend you file one side only and use the other as the rubbing surface. *NOTE*, when the correct stock *is* supplied, it may be metric — 1.5 mm instead of $\frac{1}{16}$ in., which is 3 thou. under. Check this, and if it is so, carefully file 3 thou. off each face of the top cylinder cover, where the slipper brackets fit. But check the assembly with the crosshead first, as it may give just the right clearance as it is. The main point is to keep them flat. Cut to length and square the ends with a file and then mark out for a hole at one end of each. Set your dividers to the centre distance and mark for the hole at the other end. It is more important that they all be the same length than that this dimension be spot on. Whether you drill holes one at a time or

drill one in all four and then do the other hole with the four bolted together is up to you. I did it the second way. Once done, countersink the hole at *one* end. Remove all sharp edges and set aside. Don't forget to square off the second end.

The little brackets, Part CK, should be made next. Chuck *very carefully* by the boss and face to thickness, centre with a Slocomb and drill no. 38. Whilst still in the chuck, mark off for the two other holes. Remove from the machine and spotface the no. 38 hole. Drill and tap the other holes, taking care that these are vertical. Remove burrs and then offer the castings to the cylinder cover, marking adjacent parts. File the bottom edge of the casting till it sits up square and beds on the little shoulder when the screw is tightened. You will have to file a little bevel on the inside at this edge, as I don't suppose the shoulder has a sharp corner. Assemble with the proper hex. screw and reduce the length of this if it projects into the gland space. This assembly is thus located sideways, so that the slide bars can't be 'out' when attached to the brackets. Do this now, using countersunk screws.

Screw up tight and set the bars parallel. Attach the two brackets to the cylinder and check for perpendicularity. Now prepare the upper spacers, part CJ. Chuck in the 4-jaw, offset as required, to get the hole in the right place, face the end and drill. You can save yourself a bit of work with the file by turning off part of the recess in the side if you like. Part off a shade over length. Now be careful. File the parted off end of each spacer so that it just fits between its own pair of slidebars at the *bottom* end, close to the cylinder cover. Take care to keep the end square. If you make one too small, make another—don't try to use a shim. Once you have the spacers fitting at the bottom, mark them and assemble them at the top with a 7 BA bolt and nut. Check for squareness.

Fitting Up

First, select and mark each crosshead slipper to its own pair of bars. File the faces, equally, until they are a tight slide fit between the bars. They may be a shade tighter in spots—not to worry—but if it is sloppy at one end and really tight the other the spacers may be wrong. The long ends of the slippers, Part CE, should point towards the cylinder.

If tight or sloppy in the middle and the reverse at the ends, the bars are bent. Seek out the cause of any serious variation and remedy it. Remember, at this stage a few tenths can feel like a yard! Once the two slippers are fitted each to its pair of bars, assemble the crosshead and offer it up complete. It will feel a shade tighter but so long as it slides without undue effort, leave it be. If it seems unduly

tight, check with marking blue—you will probably find the crosshead faces a little on the skew. Again, don't go looking for the odd tenth yet—it's just the extremes you must avoid.

The final trial is made with the cylinder, piston, and rod in place as well. Fit the gland (no packing) and thread the piston rod through the cover. Screw it onto the crosshead fork, tight. Fit the cover to the cylinder and insert a couple of temporary screws. Check that this lot (without the crosshead slippers) slides reasonably. If not, look for and correct the trouble. Fit the slippers and again try, over the full piston travel. It will feel tight but all is well as long as it isn't tight in spots, or the parts visibly trying to grind away their mates! Look for and correct faults at each stage and you should have no trouble. Once you are satisfied, remove both slippers and *very* slightly ease both faces of each. Just 'show them a scraper' as they say. Do this until the piston and rod assembly slide up and down without shake, and without binding—with no oil, by the way, at present.

We must now fix the bars permanently to the brackets. Detach the brackets and bars complete from the cylinder cover and douse the lower ends in a degreasant

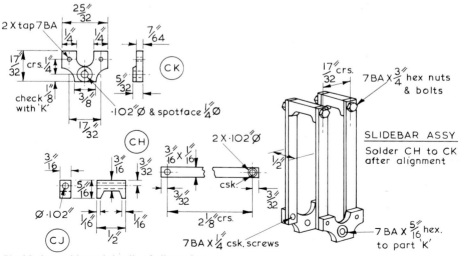

Fig 33-Assembly and details of slipper bars.

—very hot water and detergent will do— and dry them. Refit, this time taking care to avoid oil or fingermarks on the lower ends of the bars. Check alignment, as before; remember, if it isn't right now it can only be the fixing of the bars which is at fault—you may have knocked them in the process. Anyway, put the job right and then apply a blob of tinman's solder (use an active flux) to anchor each bar to the lower bracket. I found that a 75 watt electric soldering iron was adequate, but the larger the better. Remove the brackets from the cover again and apply more flux. Now solder the bars to the brackets properly, trying to avoid melting the initial blob until you have a good anchor elsewhere. Make a neat joint, then deflux thoroughly and file off any surplus, as well as any projecting screw heads or tails. You can solder up the top spacers as well, if you like; they won't have to come off again unless you have a wreck at any time.

Make a final assembly check, but I think you won't have any trouble at this stage bar a trifle of binding if a bit of solder has got part way up a bar. Before going on to the next bit, a word about assembling the connecting rod. The small end strap and cotters are assembled first, and checked that the back of the strap doesn't foul the root of the fork. If it does, file the back of the strap—it won't show. The slippers are dismantled and the bush pushed out. Insert the rod, then the bush, then the two slippers and finally the bolt. You may find a slightly tapered pilot peg a help in getting the bush into place. Oil all before assembly.

Trial Assembly of Engine

It is well worth erecting the engine and giving a trial under air or steam before going further. Pack the piston with graphited asbestos yarn and try it several times in the cylinder to make sure it is packed evenly. Also the valve and piston rod glands, which should be about half full when screwed down. Knock out the necessary gaskets—I use 'Oakenstrong' from Stuarts, but good brown paper will

do at the pressures we are working at. Oil the joints well. Make holes for the bolts with a leather punch if you have one. Put together the cylinder set, but without the slipper bars at present, and adjust the glands until they seem a bit tight. Use temporary screws in the steam chest and cover, but don't bother with the valve. Fit the slipper bars, and then attach the whole to the base with a couple of screws. See if you can wangle—and it is a bit of a wangle—the whole into the column. If not, mark where things foul and enlarge the recesses in the bottom flange of the column to clear.

Next assemble the crank, with eccentric pulley into its bearings in the entablature. Check for clearances. Attach to the top of the column and fit the rod. You may find it fouls—file away at the offending part of the column or entablature. Remove the rod and fit the eccentric and its rod, and again check and correct as need be. By the way, when you do all this, have a brass setscrew in the flywheel, not a steel grubscrew, to avoid damaging the shaft. You can now return to the cylinder set, fitting the valve and rod and tightening its gland as for the cylinder. Assemble without chest cover, and fit the connecting rod. Wangle the column on and attach it to the base with two screws. Drop on the entablature, crank, etc., and attach that also. Put on top dead centre and assemble the big end to the crankpin. There should be a slight trace of clearance between crankweb and adjacent main bearing, so scrape or file the latter if there isn't. Check for rotation. If the piston clouts either end of the cylinder you have either made a gross error in machining or the entablature isn't seating in the bottom of the spigot cavity. Have a look before shooting yourself! If the thing feels stiff, that's as it should be, but if irregularly stiff, look for the cause and correct it.

Attach the valve rod to the eccentric rod, and rotate carefully, checking that the valve doesn't foul the end of the steam chest. This established, adjust the valve position so that you get equal port

Fig 34-Initial test of the author's engine under steam.

opening at each end of the travel—don't bother with the eccentric setting yet. Now decide on the way you want the engine to rotate—the 'correct' way is clockwise looking at the flywheel, but I think this engine would be better if clockwise looking at the front. It's up to you! Set the crank on TDC, and rotate the eccentric (the 'right' way) until the valve is just about open—just a crack of port showing. Tighten the grubscrew lightly and rotate to the other dead centre—again, going the right way round. You should have exactly the same lead. If not, and the difference is slight, adjust the eccentric to get both nearly alike, but with the larger lead at top dead centre if they are different. If there is quite a bit of difference, then you should adjust the position of the valve on its spindle and try again. Unfortunately you can only move the valve half a thread, and this limits the adjustment available, so you will have to compromise. If you want really sweet slow running, let the smaller lead be just the odd couple of thou., and

let the other, at TDC, be what it comes out to be. You can make the final adjustment after she has bedded in.

Now, if you are lucky you may be able to get the steam chest cover on without taking all to pieces again—the production engine columns are slightly different from the prototype in the photographs. But I fear that you may have to take it down, and if you do, mark the accessible face of the valve rod fork with a felt pen so that if you move it accidentally you can put it back. Anyway, detach the connecting rod at the top, and the eccentric rod at the bottom, and take the job apart. Apply a 'slippery' oil into the valve chest (Colloidal graphite is my favourite, but the Molyslip type will do) and fit the cover. Reassemble, but this time put the full complement of screws in everywhere. Fit the base to the plinth if not already done—I nearly forgot that! Screw on a temporary pipe and try a whiff of air. She may need as much as 20 lb/sq. in. at first, but will rapidly speed up as things bed down. If it were to need much more you would hardly have been able to turn the wheel by hand. After about 5 minutes of running examine the following for tight spots: slidebars, eccentric strap, and the big end bush. If necessary, use a scraper on any high-spots. Start up again and this time check the main bearings. Slacken off each pair of screws in turn, and if the engine speeds up you have a slight trace of mis-alignment. If only slight, leave it, but if it is a marked change of speed file the top of the main bearing seating in the entabla-ture to correct the alignment. Crankshaft endfloat should be a bit random, as the end location is intended to be between the governor pulley and the flywheel, but if the crankweb tries to bear sideways every revolution the crankpin may be a bit out of line or the rod is twisted a bit. A little bit of this can't be helped, but if it is obtrusive you may have to take all down and correct at the seat of the trouble.

Running in

Run as much as you can whilst getting on with other jobs, once she has got down to 10 lb/sq. in. or so. You can safely run up to 1000 r.p.m., but above this she may walk about a bit—but who wants to run at that speed, anyway? It will pay you to make up the exhaust pipe at this stage, if only to avoid puddles of oil on the bench. It is not a difficult job but anneal the bend frequently as the tube is fairly stiff. I had to do this five times. Don't solder on the oval flange at this stage—leave this step until all is finished on the rest of the engine. I also recommend a steam test as soon as you can—you can't see *air* leaks! If you *DO* get a gland leak, well I am very sorry, but you may have to dismantle to get at one of the cylinder gland nuts. Or make a special spanner. It's not my fault—the engine is to scale, but the spanners are not! Incidentally, at this stage you can well put an extra ring of packing in the glands, as they will be well bedded in by now. She runs very well on steam, and even a tiny boiler will keep her ticking over for a long time. The photo in fig. 34 shows the prototype bedding down nicely at about 600 r.p.m. on quite a small spirit fired boiler; which was 'blowing off' at that!

Chapter Eight

GOVERNOR

Governor Drive Arms, Part DD
fig. 35 page 60

File one edge of the sheet material provided straight and cut the piece into four equal parts. Solder these together to make a thick lump and then file two edges square to each other. These are the reference faces for the marking out, which is done in stages as shown in fig. 36. This should be done very carefully, as this part of the engine is prominent. Once marked out, drill the no. 60 holes; pop lightly, examine, deepen the pop and examine again, then drill. If the holes then do not appear to be in the centre of the section, adjust the latter to make them so. Put a $\frac{7}{32}$ in. drill through at the centre from which the $\frac{1}{4}$ in. radius is struck (step 2) if you like, to ease the cutting out of the shape. (See page 62.)

Remove burrs and take off all surplus solder with a coarse file, and then bring the outside to shape with a fine one. Holding is not always easy, and fig. 37 shows one way of doing it. Work right up to the line—don't leave a witness but don't go beyond. Having formed the outside, cut down to the drilled hole in the middle with a couple of saw cuts to get rid of most of the material and then finish with a file. As you get towards the line keep an eye on the shape and appearance. This is more important than true dimensional accuracy —the only ones that matter are the no. 60 holes and you have these in place already. Once you are satisfied, unsolder and remove all the solder from the faces of the arms, but leave the inner faces at the straight edge tinned, where they will mate with the carrier. See fig. 38. Set aside where they won't get lost.

Governor Carrier, Part DE
fig. 35 page 61

Grip the pulley boss in the 4-jaw, chucking piece outwards, and true up the latter. If your casting has no chucking piece—and I understand that a few of them have none—face the end, machine a flatbottom recess say $\frac{1}{4}$ in. dia. and $\frac{1}{16}$ in. deep. Then solder (or braze) in a piece of $\frac{1}{4}$ in. dia. brass about $\frac{3}{8}$ in. long to serve as one. Change to the 3-jaw and hold by the machined chucking piece, pulling the outer end until it runs true, and centre it. Bring up the tailstock and take a very light cut over the edges of the vanes or ears just to bring them parallel. Rough out the pulley to a shade over diameter and machine the accessible face. Withdraw the tailstock and *very* carefully drill 2.4 mm. or no. 41. Go easy and clear chips frequently. Take from the chuck and file the length of the ears to dimension. Put a little centre in the drilled hole each end with a Slocomb held in the fingers and then set the work between centres— you will have to improvise a bit to get a carrier on. Turn the cylindrical shank to 0.187 in. dia. by micrometer—another reference dimension—and finish the pulley and the boss at each end. Part off the chucking piece, thus also facing the pulley boss.

Set up the vertical-slide and vice facing the headstock and grip the work by the machined edges of the ears—the face of these vertical and the drilled hole hori-

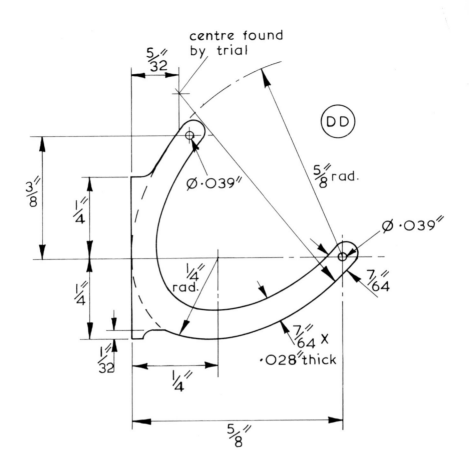

centre found by trial

$\frac{5}{32}$"

DD

$\frac{5}{8}$" rad.

Ø ·039"

Ø ·039"

$\frac{3}{8}$"

$\frac{1}{4}$"

$\frac{1}{4}$"

$\frac{1}{4}$" rad.

$\frac{7}{64}$"

$\frac{1}{32}$"

$\frac{1}{4}$"

$\frac{7}{64}$" x ·028" thick

$\frac{5}{8}$"

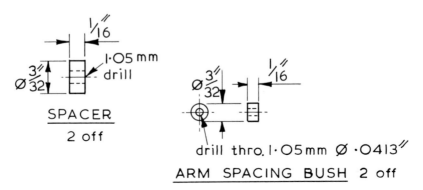

$\frac{1}{16}$"

1·05 mm drill

Ø $\frac{3}{32}$"

SPACER
2 off

$\frac{1}{16}$"

Ø $\frac{3}{32}$"

drill thro. 1·05mm Ø ·0413"

ARM SPACING BUSH 2 off

Fig 35-Governor drive arms and carrier details.

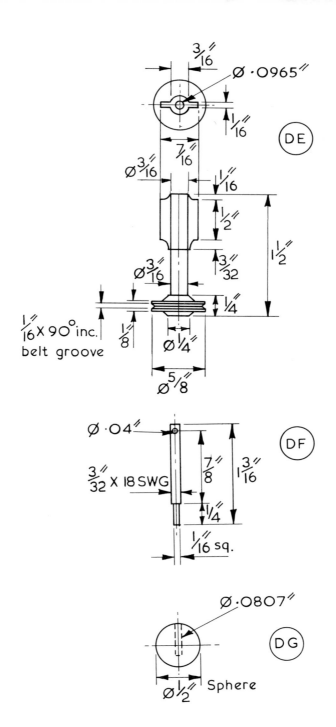

$\frac{3}{16}''$

$\varnothing \cdot 0965''$

$\frac{1}{16}''$

$\frac{7}{16}''$

$\varnothing \frac{3}{16}''$

$\frac{1}{16}''$

$\frac{1}{2}''$

$1\frac{1}{2}''$

$\frac{3}{32}''$

$\varnothing \frac{3}{16}''$

$\frac{1}{4}''$

$\frac{1}{16}'' \times 90°$ inc.
belt groove

$\frac{1}{8}''$

$\varnothing \frac{1}{4}''$

$\frac{5}{8}''$
$\varnothing$

DE

$\varnothing \cdot 04''$

$\frac{3}{32}'' \times 18$ SWG

$\frac{7}{8}''$

$1\frac{3}{16}''$

$\frac{1}{4}''$

$\frac{1}{16}''$ sq.

DF

$\varnothing \cdot 0807''$

DG

$\varnothing \frac{1}{2}''$ Sphere

61

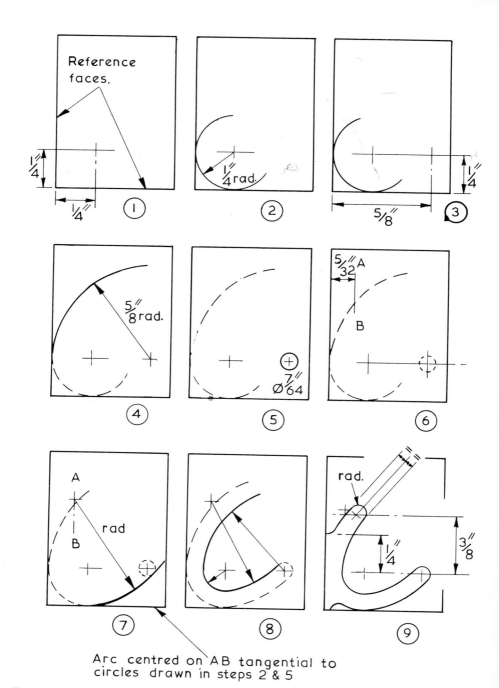

Arc centred on AB tangential to circles drawn in steps 2 & 5

Fig 36-Steps in marking out the governor drive arms.

zontal across the machine. Put a piece of $\frac{3}{32}$ in. silver steel through the hole and wangle the casting until it is truly square across the machine and the ears vertical. Set a $\frac{3}{16}$ in. endmill in the chuck (or smaller if need be to clear the vice jaws) and then engage the leadscrew hand-wheel and advance the saddle until the end of the cutter just traps a piece of paper against the $\frac{3}{16}$ in. dia. machined shank. Note the index reading. Raise the vertical-slide and advance the saddle, then adjust the former until the *side* of the endmill traps paper against the $\frac{3}{16}$ in. dia. —note the reading. Repeat for the cutter touching the opposite diameter, underneath. You now have three 'zeros' to enable you to set the cutter when machining the ears. Set the vertical-slide to one of these zeros and advance the saddle to the leadscrew handwheel zero. Then, using the latter to put on say 5 to 10 thou. cut at a time, machine across the ear till you have advanced by $\frac{1}{16}$ in.— .063 in. Repeat at the other vertical-slide zero, and the machined edge must be exactly $\frac{3}{16}$ in. wide and $\frac{1}{32}$ in. from the carrier centre. Take from the vice and saw off the unmachined part of the ear, and remove burrs. Now hold the carrier by the machined shoulder and tap it back to the vice jaws. Repeat the procedure of setting and machining on the other side as before, and the job is as true as possible. Take from the vice, and remove burrs. File off any surplus metal from the ears after the weight-carriers are soldered on.

Weight and Arm, Parts DF and DG fig. 35 page 61

The weight arm comes as a strip over-width and length. Centrepop for the holes—work as one piece—and scribe a circle therefrom to give the correct width. Drill and then file to width. Cut to the correct length, and then file the shoulders, taking care that both pieces are alike. Remove burrs and polish. Grip each ball in turn in the 3-jaw, centre and drill no. 52 —better too small than too large. Try the

Fig 37-Holding the drive arms in a small vice.

arm in the hole, and ensure that it is an easy but not sloppy fit. Mix some Araldite according to the instructions, and after degreasing, smear some on the peg and put a little down the hole. Push in the peg (make sure it doesn't push itself out again) and set aside to cure. Any surplus Araldite may be cut off after about 5 hours, thereafter leave as long as you can in a warm place. You may prefer to use solder, but take care. It isn't easy to keep

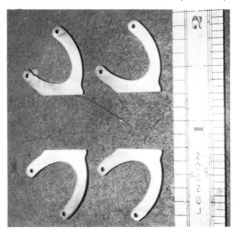

Fig 38-The finished arms, after separating and cleaning.

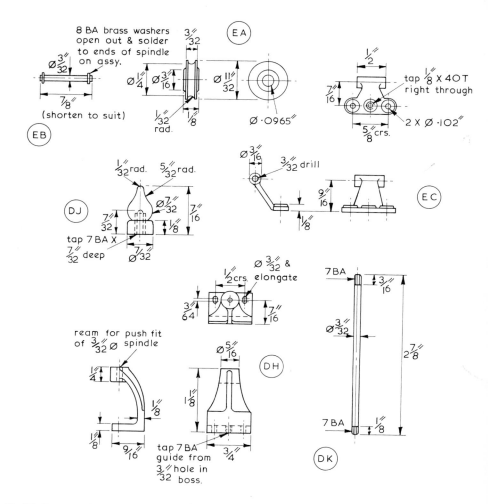

Fig 39-Governor brackets and details.

the polish on the ball if you have to clean off any splashes.

Governor Bracket, Part DH fig. 39

File the base first, and then the front face square to it—it's not worth machining either, Clean up the top of the boss to dimension and mark out and centrepop for the hole. Drill no. 42 and then enlarge to fit the material provided for the spindle—a slide fit. Put the last drill right down to make a centre, and drill the base 7 BA tapping size; tap this from the top. Turn the job upside down and mark out and drill for the holding-down bolts. Elongate these as shown and then trim the casting for paint. I don't think I need say anything about either the spindle or the nut (DJ and DK) except that the former should be brought to a high polish and the tops of the threads reduced to pass easily through the carrier, part DE.

Jockey Bracket, Pulleys and Spindle
Parts EA, EB and EC fig. 39 page 64

The jockey bracket is not a casting as suggested by the drawing, but fabricated from a rod and a piece of sheet. Chuck the piece of $\frac{3}{16}$ in. rod supplied, face the end, drill $\frac{3}{32}$ in., and part off to length. File the flat brass to shape, making the sloping tongue $\frac{1}{2}$ in. long—this will give about the correct position of the boss in relation to the base. Cut part way through (see fig. 40) and carefully bend to shape. Check the dimensions by offering to the boss, adjusting as need be. It's a bit of a job to hold the parts together when brazing, but I found that by wedging them between two bits of firebrick I managed well enough. It might pay to file a groove in the boss, and fit the end of the tongue into that. Braze all together (Easiflo no. 2) and put a touch of brazing into the saw-cut. Clean off and pickle, and then drill the holes in the base—I suggest no. 35 for the outer ones. Finally, solder 7BA

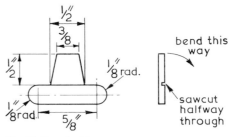

Fig 40-Detail of jockey bracket, part EC.

brass washers over the holes to form the little seatings for the nuts and lubricator.

The little pulleys (EA) call for no comment except to emphasise that the profile of the groove should be half-round, not a vee. The pulleys should be an easy running fit on the spindle (EB) and you should lightly countersink the ends of the holes. The spindle length is taken from the job—assemble the pulleys and washers, cut to length, and allow a bit for riveting or soldering. Leave a little side

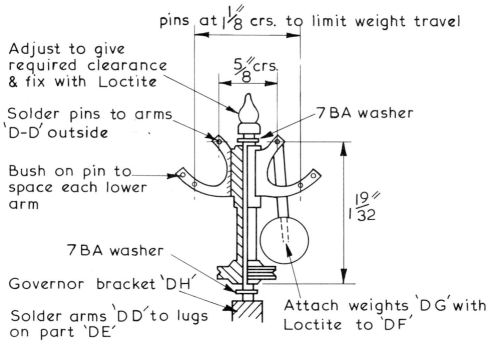

Fig 41-Assembly of governor. (Note: The vertical clearances are exaggerated.)

Fig 42-Close-up photograph of governor, showing run of drive cord.

clearance for the pulleys—not much. Don't assemble until you have tried it on the engine as the bracket may need bending a little.

Erecting the Governor

The general arrangement is shown in fig. 41. The biggest difficulty is soldering the drive arms to the carrier, and I recommend that you assemble these with a clamp, drill very small holes through, and fit rivets—later to be filed off—to hold all whilst soldering. To get the arms correctly aligned, slip a piece of pinwire through the upper holes, and make up the joint at the lower hole in each pair. The latter calls for a bush, part DD, for which you simply chuck a piece of $\frac{3}{32}$ in. brass, drill no. 59, and part off $\frac{1}{16}$ in. wide. Poke a bit of wire through the hole when parting off so that you don't lose it. Tin the surfaces, assemble with the rivets I mentioned earlier, check the position, and

solder up. It is important that the two upper holes be at the correct height and distance from the centre of the carrier, so check this carefully. This having been done, file round the junction of the arms and carrier, making any sharp corners into little radii, and generally clean up. The weights can then be attached, the pin being lightly riveted over outside the arms and given a touch of solder.

Fitting the Brackets

The governor bracket must be fitted to the machined pad on the entablature so that the spindle stands dead central. Do this by eye—it is what it looks like that matters—and locate it so that it doesn't project beyond the front of the casting; $\frac{1}{64}$ in. back is OK. Mark through the holes, and then drill and tap for the studs. You can then adjust for position with the enlongated holes after the decorative panels are fitted—the elongation is not for belt tensioning. The jockey fits above the flywheel main bearing; fit the drive pulley so that it locates the shaft endways, slide on the main bearing, and then the bracket on top.

Fit the jockeys, and then adjust the position—bend if need be—until the cord will run straight onto the little pulleys. (See fig. 42). If the drive pulley fouls the bracket either bend or file the latter to clear. I had to bend up to clear the pulley, and then the top boss forward to get the cord straight. I used a linen fishing line, dewaxed, as a drive cord, but a spring band as supplied for the MAMOD toy steam engine serves very well. Don't use nylon—it slips. A MECCANO spring drive cord will also serve, but rubber cords perish very quickly.

Chapter Nine

FINAL ERECTION

Decorative Panels, Parts F and G fig. 43

Leave these until all other work, test running, and perhaps even painting is done. The Araldite adhesive is very strong, but it is a bit brittle, and a chance knock whilst doing something else could loosen the joint. Note: these panels are die-cast in a soft metal, so handle with care, especially when clamping or cutting off any surplus Araldite after assembly.

However, you must dismantle a bit. Take off the governor, undo the eccentric and big end bearings, and then lift out the shaft complete. Detach the entablature from the column, and then thoroughly degrease it. (Boiling in detergent solution is perhaps the best way, but if you have painted it, douse in CTC or Trichlorethylene.) With a coarse file abrade the machined surfaces which carry the panels, and again degrease. Take the front plate and file the back surface flat—if it needs it —and then offer this to the entablature. Mark for the two little rebates and file these so that the dimension shown as $2\frac{3}{16}$ in. is a trifle less than the width of the entablature and the thickness of the rebate is $\frac{1}{16}$ in. Finish the edges to dimension, and do all other work— polishing etc.—needed but don't shorten at the ends.

Take each side panel in turn and treat likewise from the rebate. For the present these will project $\frac{3}{32}$ in. or so beyond the front of the entablature. Mark each to its mating side (they will be 'handed' when you've finished) and then file the front edge of each until, when the front plate is also offered up, it meets both the front plate and the entablature. Try and get the join as unobtrusive as possible. Collect together all your clamps, and have a rehearsal as to how you are going to clamp all up whilst assembling. If you can put them in a suitable position you could fit little dowels to help hold the job if you like. I prefer not to; if the drill wanders there is little chance of correcting it. Have several tries until you are satisfied that you can clamp up without any hesitation. Thoroughly abrade the backs of the plates. Note: Use silicon carbide grade 100 or 150, or even sandpaper, rather than emery cloth. You want sharp scratches to get the best adhesion.

Degrease everything *again*, and thereafter avoid touching the surfaces with your fingers. Mix the Araldite as instructed —it helps if you warm the tubes for a few minutes first—and then coat the front face of the entablature, a thin even film. Offer up the front plate and clamp up lightly. Repeat for both sideplates, working quickly but carefully, but in this case apply a little Araldite to the rebates also. Have a good look at the job and make sure nothing is askew, adjusting if need be. If all is in order, tighten up the clamps just enough to ensure that nothing will shift of its own accord, but don't have so much pressure you squeeze out all the 'glue' and get a dry joint, or worse, mark the metal. Wipe off surplus, and make sure no Araldite gets on the clamps. Now leave (a) at room temperature for 6 hours or (b) on a storage heater for $1\frac{1}{2}$ hours, and then shave off any surplus Araldite

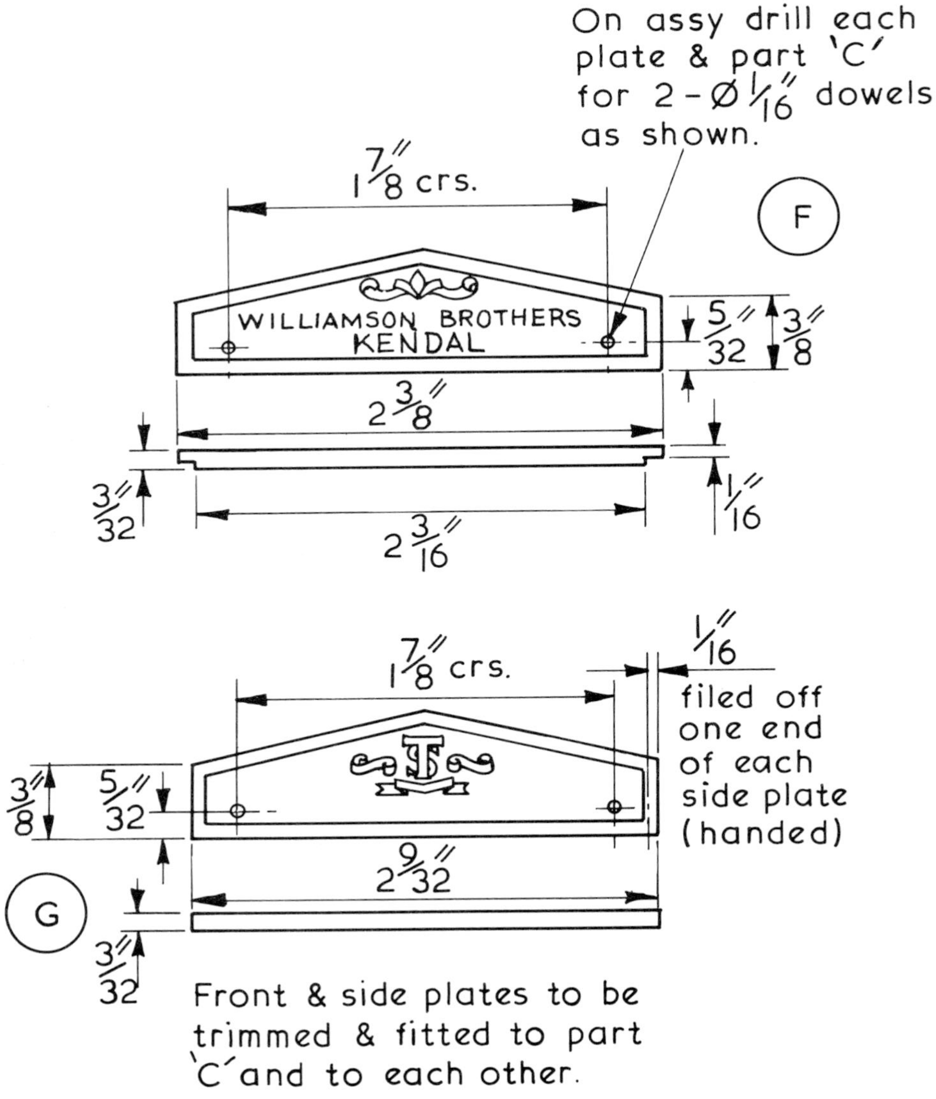

On assy drill each plate & part 'C' for 2 - Ø $\frac{1}{16}$" dowels as shown.

WILLIAMSON BROTHERS
KENDAL

$1\frac{7}{8}$ crs.

$\frac{5}{32}$" $\frac{3}{8}$"

$2\frac{3}{8}$"

$\frac{3}{32}$"

$2\frac{3}{16}$"

$\frac{1}{16}$"

F

$1\frac{7}{8}$" crs.

$\frac{1}{16}$" filed off one end of each side plate (handed)

$\frac{3}{8}$" $\frac{5}{32}$"

$2\frac{9}{32}$"

G

$\frac{3}{32}$"

Front & side plates to be trimmed & fitted to part 'C' and to each other.

Fig 43-Details of front and side plates to entablature. See text, page 67, regarding the dowels.

that remains. You can remove the clamps now, and if you are dissatisfied with the job, take it apart, clean down and start again. Otherwise, leave in a warm place to cure—overnight on a radiator, longer elsewhere. Finally, trim off any overlapping edges and repolish if need be.

Painting Most people have their own ideas about this, but it is a repeated criticism of models shown at the ME and at Local exhibitions, so perhaps a few words will not be out of place. The main snag is that 'scale paint' is a very difficult concept—a brush-mark on a model would be a groove you could shove your fist into on the full-size engine! So you must aim for the very finest finish you can manage, and this means only one thing—CARE! Unfortunately by the time you get to this part of the job you want to get the thing running, and it is hard to exercise patience. But you must *be* patient, and give the job the time it deserves.

First, colour. This is entirely a matter of personal taste, but far too many people think that 'engine green' (whatever that may be) is 'de rigueur'. Not so. Many early engines were plain black, probably because this was the only pigment that would stand the heat, oil, and dirt. And it doesn't show oil-stains! But one very famous firm used maroon, another royal blue and agricultural machinery has been painted red-and-yellow for a very long time. The big blowing engines at the old ironworks were a rich brown, and I have seen waterworks engines with scroll-work on the ornamental castings picked out in gold! The one thing to avoid is 'pastel shades'—colours should be deep and rich—and if you *do* use green make sure that it isn't the sickly colour found in so many paint sets. You may well have to mix your own to get it right. Small details, like crankwebs, crossheads, etc, were usually picked out in bright red, and even quite early on, pipework was coloured according to duty. So, you have a wide choice, and the guiding principle should be to choose a colour that looks right *TO YOU*, and never mind what the judges say!

Quality in paintwork depends entirely on the 'underwork' and the topcoat just adds the finishing touch. Start by really thorough removal of odd knobs and lumps, and then abrade the surface, especially on machined parts. Next degrease thoroughly, and give a coat of primer. I use cellulose for all underwork, irrespective of the top coat, because it's quick—you can use an airbrush, paint gun, or the aerosol 'Auto-touchup' for much of it and coats dry very quickly. Apply a second priming coat (preferably a different colour) as soon as the first is touch dry and then leave for twice as long as the makers recommend, to set. The next coat can conveniently be 'primer-filler' or 'Primer-surfacer', which is brushed on. Let this set overnight and then rub down with soft-back (the softest you can get) wet-and-dry silicon carbide paper grade 320 on plain surfaces, 360 on complicated ones with plenty of water. (Note, on a machined surface this coat can be omitted). Rub down until the bare metal shows almost all over—the object is to fill the hollows in the surface. Reprime as before, and apply another coat of primer-surfacer; *but* if there are any really deep hollows, fill these with 'cellulose putty', applied thinly with an old table-knife. Allow to set overnight, and then rub down again, to the metal, using 360 or 420 grade this time. Spray on another even coat of primer, and when this has set, examine under a strong light. If there are any hollows still showing, repeat the process, but on a small engine like this you should have filled them all by now. If all is well, spray on two or three thin, even, coats of primer, taking care to avoid runs and blobs. If you have them, use alternate colours—grey and red. *LIGHTLY* rub down this coat until the next colour begins to show—use the worn 420 paper from the previous work, and add some soap to the water. If there is the odd run, rub this flush to the surface, exercising great care. If by any mischance you go right through to the metal anywhere, reprime that part and very lightly rub down again.

Now apply a thin coat of colour; if you are using a spray, several thin coats are desirable; if you intend to use a brushing enamel as the final coat, use a cellulose spray almost the same colour. Build up the thickness gradually, and then rub down with no. 420 (or even no. 600 on a surface like the column) with soap in the water. Matt the surface all over, but stop as soon as the primer begins to show. Now apply a full coat of final colour—spray or brush as you wish. Rub this down with no. 600 and soapy water, very VERY lightly, but make sure that the surface is matt all over.

Fig 44-The finished engine.

Provided there are no cracks in the abrasive coating it pays to use the paper you have already taken the edge off in previous work. You may now apply the final coat, and I need not remind you that cleanliness—of the brush (if used), the job, the workplace, everything—is essential. Have a dust-free box handy to drop over the job as soon as painting is finished, and cover it to keep flies and debris from settling on it.

For this final coat use either: spray cellulose, or a *SLOW*-drying enamel applied by brush. *Don't* try to brush on a quick-setting paint—you must have time to go round and pick up any runs, or remove the odd loose brush-hair. If using a spray, better to put on two thinnish coats than one thick one. Leave the job until the paint is set (paint a test surface at the same time) and then for twice as long again before handling.

For small details—say the crosshead—this lengthy procedure is obviously impracticable. For such parts I use either the brushing cellulose or the little Humbrol tinlets of paint. Attach the parts to a rod or wire to hold them, and have a hook handy to hang them up in as dust-free a place as you can find. I usually spray on a primer, but with the Humbrol paint this doesn't seem to be essential on well-abraded surfaces. If you need to rub down, then use the finest steel wool instead of wet paper, but make sure no fragments of the wool remain adhering to the surface. Where one colour has to go on top of another (you may wish to pick out the wheel spoke flanges, for example) then I recommend that you don't use cellulose for the second colour—use an oil enamel; there is a risk of the cellulose vehicle softening the paint underneath with unpleasant results.

Humbrol and similar enamels are self-surfacing, but all cellulose needs a final polish to get the best finish. You can get proper rubbing down compound, but it is a bit fierce, and I use Brasso for this job. Even then you must be careful not to polish right through, especially on corners. Rub in straight lines using a new yellow duster or old wincyette (they make kids' pyjamas of it) not cheesecloth or other harsh cotton. Exert very slight pressure and be as gentle as you can, and you will find it comes up like a mirror. Most important, however, *leave this operation until the paint is a week old.* You *CAN* do it before, but it pays dividends to be patient.

Final Assembly

Retouch any paintwork that has been damaged and have a look round for any signs of excessive wear or slack, especially on the eccentric strap. As a rule, if an engine has run 6 hours or so at 500–1000 r.p.m. (that's a third of a million revs. or so!) any 'bright marks'—e.g. on the slide-bars—are only a fraction of a thou. and finger pressure will correct any misalignment. But glands may need attention, as they bed down quite a lot in the early stages. Remove any 'black oil', graphite etc., and replace with proper steam oil. I see I have forgotten about the exhaust and steam flanges. They are a simple job, though, and I expect you have already made them. Don't forget to make a little paper joint for the oval exhaust flange. If you need union connections I suggest you terminate both 'permanent' pipes with a flange, and have a flange-to-union adaptor for working. Unions look all wrong when the engine is on show, whether at home or in an exhibition!

On my own engine I fitted the Stuart Turner (fig. 155) combined lubricator and stopvalve. This 'looks right' if you alter the plastic handwheel and certainly improves the engine's running. I think that's all. Fig. 44 shows the finished engine, and though it may differ a little from yours (some improvements have been made in the 'production' casting sets) I think you will agree that she looks very well. I hope you have had fun making it, and that yours runs as well as mine.

Chapter Ten

The engine is elegant enough to make a 'Glass Case' model, and it can, of course, be run on compressed air, but after all, a steam engine is supposed to run on steam. Many builders will already have a test boiler in the shop—I have one myself—but all too often this is overlarge for the duty and it is not only disconcerting for visitors but also bad for machinery to have the safety valve blowing all the time. The little boiler shown in the photo (fig. 45) is one I built in a few hours one evening specifically to run the 'Williamson'. As far as possible it is about the right size and shape, typical of the simple vertical boilers which might have been supplied at the time the original engine was in production.

I have used spirit firing, as this is convenient, and with the lamp shown later will raise steam in about 6 minutes, and run the engine for nearly half-an-hour at one filling with water. Design pressure is 30lb/sq.in., but I set my safety valve at 20, which is quite enough for the purpose in mind. In the event, the boiler has been very useful for other purposes. I doubt if it would be adequate to 'run in' a newly constructed engine of more than say $\frac{3}{4}$ in. bore, but it *has* driven a 1 in. × 2 in. stroke beam engine at a stately 75 r.p.m. when well warmed through first.

I must confess that the boiler as built, and as shown in the photos, has some faults. The sleeve joining the chimney to the top of the centre flue is far too heavy—in any case, for a 'proper' model it should have a flange, real or dummy, here. The water level test cocks were made from commercial drain cocks with a length of pipe soldered in; these extensions are too long. The dummy firehole door could have been better made—it isn't thick enough in material, and should be a trifle larger; there is a note on this matter later. Finally, I made the great mistake of painting it with shiny paint! *Never* found on boilers! Red-lead colour is correct for a boiler of this type (or black) and it should be matt. The spray primer used on the engine, say 3 coats, will make yours look better, the firehole door and chimney being, of course, matt black.

Construction has been made as simple as possible. Top and bottom end-plates are identical apart from the bushes in the top for such fittings as are decided upon. The flue is a piece of ordinary commercial copper water pipe, and the shell may be either copper tube 20s.w.g., or a rolled and jointed sheet of 18s.w.g., the increase in thickness to allow for the 'efficiency' of the joint. The firebox is a rolled sheet of copper or brass, and the base on which it sits is made of ternplate (lead-coated steel) but could be tinplate if desired. Two bushes are shown for water level test-cocks; the builder can fit a water gauge of the usual type, but I used test cocks, as normal on small boilers in those days. However, if you *do* fit test cocks they should be off centre to each other, not one above the other as I made them; this prevents the lower cock from being flooded with hot steam or water when the upper one is opened. A 'spiral retarder' is fitted in the flue; the boiler works well enough without it, but the draught is a bit fierce for a spirit lamp, and flue gases are rather hot. The retarder slightly improves the heat transfer. The lamp is a simple 'pot' with three wicks and can be refilled at least once without refilling the boiler. Naturally, constructors can add feed-clacks, pressure gauge, and so on if desired, to add to the realism, even if not strictly necessary! Fig. 46 shows the sectional arrangement.

Formers for beating out the boiler ends are shown in fig. 47. The sheet should be cut to a circle and annealed, after which a $\frac{5}{8}$ in. dia. hole is cut in the centre. The edges of this may be beaten down over a jig shown in the sketch to form the flange round the flue tube but I just used the rounded edge of a small bench anvil. The hole should be bell mouthed until the flue tube *just* enters. The plate can then be set on the flanging jig, the first step being to beat down the conical body profile. After re-annealing, beat down the circumference to form the flange. Do this a little at a time, don't be in a hurry or it will cockle. I found it necessary to re-anneal four times in this operation. When finished it should only just go into the shell tube—better if it won't quite go in. *Note:* If you are forming the barrel from sheet, don't make the formers until after the barrel is brazed up. Then take the dimensions from this.

Now take the flanged plate and on a soft surface—a bag of dry sand, for example—and with a wooden or leather ball-ended mallet, slightly dome the cone so that it adopts a curved rather than a conical shape. Finally, re-flange if need be to fit the shell-tube. Take care that there are no gaps in the fitting. The holes for the bushes can now be made. I have a punch for such jobs, but a drill will do provided you have a metal backing under the sheet. It pays to use a drill one size down and enlarge with a taper reamer or the tang of a file. The holes should be a good tight fit for the bushes.

If you use tube for the shell, no problem; all you have to do is to drill the

(Note: This photo and fig 53 show the boiler as first constructed. The test cocks and flue sleeve should be corrected as described in the text.)

holes for the two bushes and square the ends up with a file. If you roll-and-joint a piece of plate, then you must remember to joggle the joint at the ends so that the flanged endplates fit properly. In either case you may care to set up a vertical row of dummy rivets to 'make it look proper'! $\frac{1}{16}$ in. rivets at $\frac{3}{16}$ in. or $\frac{1}{4}$ in.

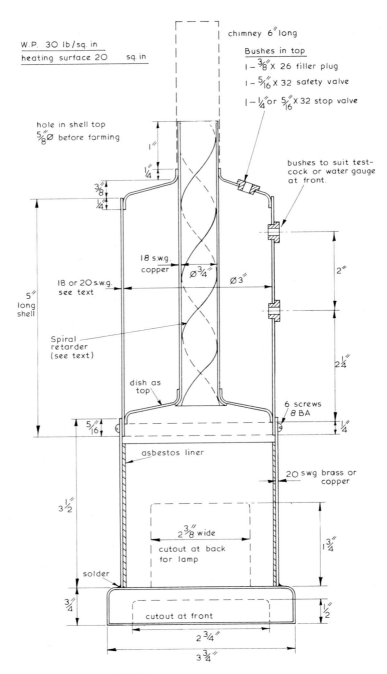

W.P. 30 lb/sq. in

heating surface 20 sq. in

chimney 6" long

Bushes in top

1 – $\frac{3}{8}$" x 26 filler plug

1 – $\frac{5}{16}$" x 32 safety valve

1 – $\frac{1}{4}$" or $\frac{5}{16}$" x 32 stop valve

bushes to suit test-cock or water gauge at front.

hole in shell top $\frac{5}{8}$" Ø before forming

1"

$\frac{1}{4}$"

$\frac{3}{8}$"

$\frac{1}{4}$"

18 s.w.g. copper

Ø $\frac{3}{4}$"

Ø 3"

2"

18 or 20 s.w.g. see text

5" long shell

Spiral retarder (see text)

dish as top

$\frac{5}{16}$"

6 screws 8 BA

$\frac{1}{4}$"

2$\frac{1}{4}$"

asbestos liner

20 swg brass or copper

3$\frac{1}{2}$"

2$\frac{3}{8}$" wide

cutout at back for lamp

1$\frac{3}{4}$"

solder

$\frac{3}{4}$"

cutout at front

$\frac{1}{2}$"

2$\frac{3}{4}$"

3$\frac{3}{4}$"

Fig 46-Sectional arrangement of boiler.

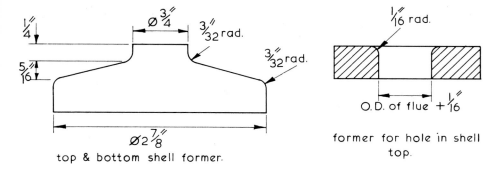

top & bottom shell former.

former for hole in shell top.

Fig 47-Formers for boiler endplates.

pitch would be about right. Incidentally, you can avoid the joggle at the joint if you use an external butt-strap at the joint, but if you do, don't forget that you should have a *double* row of said dummy rivets!

The lower end of the flue $\frac{3}{4}$ in. dia. × 6 in. long is bell-mouthed; simply anneal the end, and beat round with a small ball-pein hammer until it fits snugly to the lower plate. Make the bushes next. These are shown in fig. 48. Don't use brass; though it should be satisfactory at the pressure there is always a risk of burning such thin sections if you are too enthusiastic with the blow-lamp. You may well have fittings available (my own safety-valve is the spare from a very ancient 'toy' steam engine!) in which case

make the bushes to suit. Note that threads are not the usual 40 t.p.i.; I dislike these for boiler fittings, and have specified 32, except for the filler plug, which is 26 t.p.i., as this one has to be undone frequently. Incidentally there is no reason why you should not enlarge the safety-valve body and use this as a filler—it has the advantage that you inspect the valve spring every time you use the boiler! The bushes are a very simple turning job which I don't think I need detail, but simply remind you to take care that the thread is square to the bedding face, and to put a countersink one thread deep at this end of the hole after threading.

Brazing up—I have warned you that if you are using a rolled shell it pays to braze

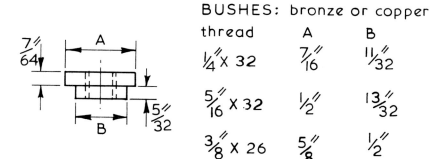

BUSHES: bronze or copper

thread	A	B
$\frac{1}{4}$" X 32	$\frac{7}{16}$"	$\frac{11}{32}$"
$\frac{5}{16}$" X 32	$\frac{1}{2}$"	$\frac{13}{32}$"
$\frac{3}{8}$" X 26	$\frac{5}{8}$"	$\frac{1}{2}$"

Fig 48-Bushes for boiler fittings.

this up before making the end-plates, so that any errors can be allowed for in the flanging. It isn't easy to get both ends the same diameter. Braze up the seam using C4 alloy if you have it, otherwise Easyflo no. 2. Heat the job from the inside of the tube, of course. If you have done the usual and fitted a few rivets to hold all together whilst heating, these should 'take' satisfactorily whilst brazing the strap or overlap, but if you have put in a row or rows of dummy rivets you may care to touch each of these with Easyflo no. 2 after the main brazing but before the job cools. Pickle, or just quench from black in cold water and wire-brush off any flux residue, and then make your end-plates to suit. (After restoring the shape to cylindrical!) The next step is to braze the flue into the bottom plate, again using the higher-melting-point alloy if available. Care is needed here to see that the flue stays square with the plate whilst heating, and you will have to rig up fire-brick packing to do this. Pickle this bit, too.

You can now assemble the parts, well fluxing each joint, including those already brazed, and fitting the bushes. The lower flange should be at least $\frac{1}{4}$ in. up inside the tube. Stand the boiler upright, but *not* on a piece of asbestos. If you do, then liquid flux may cause some of this to adhere to the metal, and result in a dud joint. I use 'fosalsil' brick ('folsain' is another similar) which is an insulating brick used in furnace construction, for this sort of job. Start with the joint of the flue to the top plate. Use Easyflo no. 2 and feed well to make sure the material fills the joint. Then do each of the top bushes, and finally the shell-to-top flange. Heat quickly, to avoid prolonged heating of the flux, which will decompose it, but don't heat to a higher temperature than needed to melt the solder.

Now transfer the flame to the boiler side, turn the boiler sideways so that these bushes are uppermost, and braze these. Finally, turn all upside down and braze up the lower flange joint. Now, it may well be that when you come to this part you find the flux all black, and it won't clarify; the brazing alloy won't take either. This is due to prolonged heating and the remedy is simple. Pickle the job, wash well inside and out with water, reflux, and deal with the bottom flange separately. You may decide to do this anyway, but nevertheless you must still flux the bottom joint for the first heat, to keep the metal clean.

The final pickle should be followed by a good swill in warm water, after which the joints should be inspected, first visually, and then by fitting plugs and water testing. 50 or 60 lb/sq.in. is quite adequate for this little chap. As it comes from the brazing the copper is all annealed and soft, and very easy to bend (e.g. when fitting to the firebox). If, when you water test, you apply and release the test pressure half-a-dozen times you will workharden the copper a bit, and improve its strength. This is more effective than leaving the test pressure on for half-an-hour or so. Any small leaks may be peened over, but 'gaps' should be filled with brazing alloy; I don't recommend soft-solder caulking other than on rivet-heads, and even then don't like it, as it means you can never do any more brazing on the job afterwards.

Firebox fig. 46 page 74

The firebox is a very simple piece of work. Make a paper template to fit the shell, cut out your piece of sheet—allowing for a lap joint—and roll it round a bit of stuff (even the boiler shell if you must!) until it is a good close fit. Note if you have put dummy rivets in the shell you should do the same on this part, *but* put the rivets so that they don't line up with those in the shell proper. Also, you should allow for a ring of rivets on the circumferential joint, too.

Once you have the joint made—it needn't be brazed, rivets alone will do—get it quite round, and then make the base. Fig. 49 shows the developed shape of the sheet, on which note that I have shown a small hole at each fold corner. This will give you both an easier cut and a better

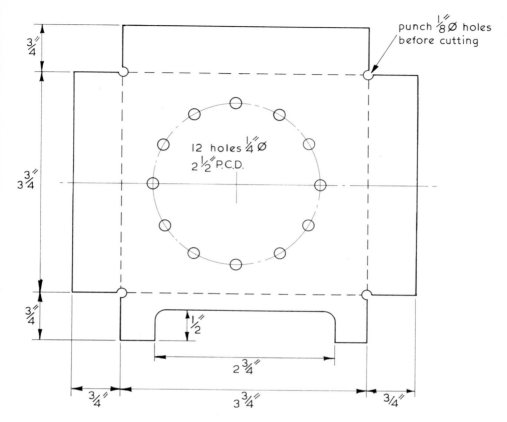

punch $\frac{1}{8}\emptyset$ holes before cutting

12 holes $\frac{1}{4}"\emptyset$
$2\frac{1}{2}"$ P.C.D.

$\frac{3}{4}"$

$3\frac{3}{4}"$

$\frac{3}{4}"$

$\frac{1}{2}"$

$2\frac{3}{4}"$

$\frac{3}{4}"$ $3\frac{3}{4}"$ $\frac{3}{4}"$

Fold on dotted lines and solder corners inside.

Fig 49-Boiler base.

fold. The 12 holes are air-holes for the burner—not essential as enough should come through the hole in the firebox. Fold it up, and run a fillet of soft solder inside each corner. Twist it by hand until it sits flat. Now take your dividers and scribe a circle on the top a trifle under the diameter of the firebox. Use this to locate the latter, and secure it with four hefty blobs of solder. This having been done, run your soldering iron all round the joint. This will stiffen up the relatively thin firebox no end. Note, if you are using brass it may be necessary to tin the joint first, but this shouldn't be necessary with tinplate. You can now try the base for flat-sitting, and once you have adjusted this I suggest you file off a few thou. in the middle of each side, so that it sits only on the corners.

Mark out for the 6 fixing screws—I used 8BA but the size isn't important—and after drilling or punching the firebox offer up the shell so that the test cocks are at the front, but the lamp-hole is at the back. (The rectangular hole in the base is supposed to be the ashpit door, so this should be at the front, too). Drill and tap the shell, and assemble. The asbestos liner is a refinement, saves the paint! All you have to do is to cut it out, using the paper template you have thrown away, wet it, and mould it to shape. In view of the current apprehension about asbestos it is worth suggesting that you cut it wet, I think.

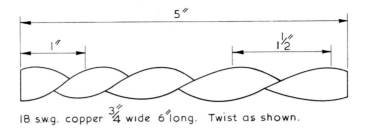

18 s.w.g. copper $\frac{3}{4}$" wide 6" long. Twist as shown.

Fig 50-Spiral retarder, to fit boiler flue.

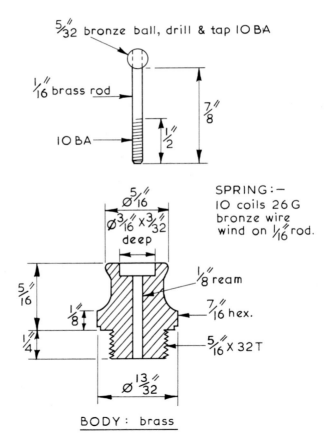

$\frac{5}{32}$" bronze ball, drill & tap 10 BA

$\frac{1}{16}$" brass rod

10 BA

$\frac{7}{8}$"

$\frac{1}{2}$"

$\emptyset\frac{5}{16}$"

$\emptyset\frac{3}{16}$" x $\frac{3}{32}$" deep

SPRING:—
10 coils 26 G
bronze wire
wind on $\frac{1}{16}$" rod.

$\frac{1}{8}$" ream

$\frac{7}{16}$" hex.

$\frac{5}{16}$" X 32 T

$\frac{5}{16}$"

$\frac{1}{8}$"

$\frac{1}{4}$"

$\emptyset\frac{13}{32}$"

BODY : brass

Fig 51-Safety valve details.

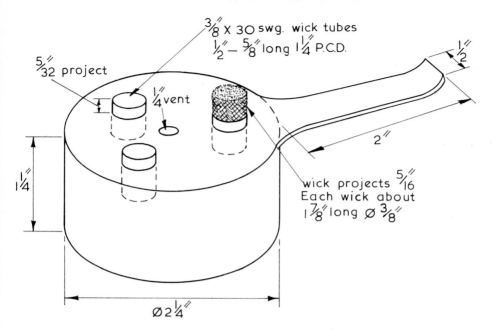

$\frac{5}{32}$ project

$\frac{3}{8}$ X 30 swg. wick tubes $\frac{1}{2} - \frac{5}{8}$ long $1\frac{1}{4}$ P.C.D.

$\frac{1}{4}$ vent

$\frac{1}{2}$

$2''$

$1\frac{1}{4}$

wick projects $\frac{5}{16}$
Each wick about
$1\frac{7}{8}$ long Ø $\frac{3}{8}$

Ø $2\frac{1}{4}''$

Fig 52-Burner for methylated spirit.

The 'Spiral Retarder' (it is really a Helix, as Prof Chaddock will hasten to remind me!) is easily made (fig. 50). Cut a strip of thin copper (anything from 28 to 22 gauge will do) a tight fit to the tube, and twist it so that the coils are about 1 in. pitch at the bottom and $1\frac{1}{2}$ in. at the top. Don't fuss over this—'about right' is spot on in this case! 'Wind' this into the flue, and to prevent it from falling right through just twist the top corners with a pair of pliers so that these corners rest on the top of the tube.

The only fitting I have detailed is the safety-valve, fig. 51, but any sort will do. A 'proper' one would have a dead weight, for the period, but these don't scale very well; alternatively, those with two columns and an exposed spring look quite well. The one I have shown is a well-proven 'LBSC' design but slightly enlarged in the body. Again, a simple turning job, the only point to note being to form the seat for the ball by tapping a *hard steel* ball into the recess, *not* the bronze one! For the stop-valve I have an ordinary union plug-cock for the time

being, to be replaced by a globe valve when I have time to make one. The filler plug is a plain hexagon head plug, and the test-cocks for water level are cylinder drain cocks, with a rather longer 'bib' of $\frac{3}{32}$ in. copper tube soldered onto the nose. These bibs are rather too long on the boiler shown in the photo, and have now been shortened.

The lamp is shown in fig. 52. This could be a bit larger with no harm done, and can be made up of either tube or sheet, but use the thinnest material you can find, especially for the top plate and the wick-tubes. I always braze my spirit lamps, but I think I may be over-cautious, as most commercial lamps are soft-soldered. You may have to make the lamp to fit such wick as you can get these days; if so, aim at a total wick area of about a third of a square inch—$3 \times \frac{3}{8}$ in. dia., or $4 \times \frac{5}{16}$ in. dia., etc. Those who live in the country may be able to get round wick still, and in some places old-fashioned ironmongers or agricultural engineers may have some 'engine starting wick' left in stock. Incidentally, there is

Fig 53-The engine and boiler. (Matt paint would suit the boiler better.)

no law that insists that wicks shall be round, and you could use a couple of ¾ in. flat wicks if you wish. Just as good, but the wick-tube is a bit more difficult to make.

When you have finished the boiler, and made a chimney with a sleeve to fit over the projecting flue (or, if you are a purist, with a flange connection!) you will think it looks tall for its diameter. Quite. This is because you haven't got a firehole door! What I did was to cut out a bit of scrap brass (actually one of the bits cut out of the firebox) to look like one, and attach it more or less in the right place, just as the old German makers of 'toy' steam engines used to do. It makes quite a difference, though I wish I had taken a bit more care over the detail of it. Never mind—a job for the future, to while away an idle moment! Such firehole doors were of many shapes—oval, round, square with round tops etc.—so you can devise your own. However, don't forget that even on the smallest boiler they had to admit a shovel, so don't go below a scale 16 in. × 12 in. for the door; say $1\frac{3}{8}$ in. × 1 in. It would be very rare to find such a boiler (in the 1860's) lagged in any way, still less with wood staves, though you might find the cost-conscious farmer wrapping it with sack-cloth or hessian tied on with baling wire. So I suggest you paint it to your choice and leave it at that. Fig. 53 shows the finished boiler, complete with engine. The sleeve joint in the flue is far too heavy and obtrusive and I hope you will make a better job of yours!

Finally—superheater! I haven't shown one, as for the purpose it isn't really essential. However, dry steam is certainly worth having, and if you feel inclined you can deal with this in one of several ways.

The easiest is to thread a pipe down the flue round the retarder, and lead the steam pipe out through the lamp-hole. Just cut a slot in the flue extension to let the pipe in. You would get rather too much superheat if you brazed the pipe to the retarder! Alternatively, you can make a hole in the bottom plate, thrust a steam pipe through this to within $\frac{1}{8}$ in. of the boiler top, braze it in place, and arrange a double coil of pipe in way of the flame in the firebox. In both cases you should use slightly larger and thicker wall pipe than usual, to allow for wastage of the metal. Alternatively, (and this is what I do) when you want superheat just lead a pipe down from the stop valve, into the firehole to a double coil, and out to the engine. Just one word of warning, how-ever; *don't* fit a superheater when driving an engine with no proper cylinder lub-ricator or you may run into trouble with a scored cylinder bore.

So, there it is; an easy boiler to make, that will provide adequate steam for the 'WILLIAMSON', run small engines quite fast, and larger ones at a nice stately speed. Don't expect everything from it, though; As I said at the beginning of the chapter I doubt if it would break away a *new* engine above ¾ in. bore and for such you would have to run in first either on your bigger boiler or on compressed air.

I should emphasise that this little fellow is—or was—designed for quick and easy construction yet with an appearance which would 'go' with the 'Williamson' Engine. Those wishing to make a more elaborate boiler should refer to 'The Model Engineer' for February 18th 1977 where a suitable, though perhaps too 'modern' design will be found.

APPENDIX

Index to figure nos. and Part Letters of Components

	Part Letters	Fig. no.
Governor, Drive Arms	DD	35–36
Governor, Drive Pulley	ED	22
Governor, Jockey Pulley	EA	39
Governor, Spindle	DK	39
Governor, Weights	DG	35
Jockey Pulleys	EA	39
Jockey Pulley Bracket	EC	39–40
Lamp, boiler	—	52
Main Bearing	D	9
Panels, decorative	FG	43
Plinth	E	5
Pin, Eccentric Rod	AJ	31
Piston	BD	12
Piston Rod	BE	12
Port setting-out coordinates	—	15
Safety Valve	—	51
Side and Front Plates	G	43
Slide Bars	CH	33
Slide Bars, Bracket (Anchor block)	CK	33
Slide Bars, Spacer	CJ	33
Slide Valve	AE	31
Slide Valve, Nut	AH	31
Spindle, Governor	DK	39
Steam Chest	AB	32 and 18
Valve	AE	31
Valve, Chest	AB	18
Valve, Rod	AG	31
Weight, Governor	DG	35

Descriptive and constructional matter will be found on the pages adjacent to the above figure numbers.

List of Castings and Materials Supplied to Build the 'Williamson' Engine
(Courtesy of STUART TURNER Ltd.)

Part Letter	Description	Material Supplied	No. of pieces
A	Column	Al. Alloy Casting	1
B	Base	Cast Iron	1
C	Entablature	Cast Iron	1
D	Bearings, Crankshaft	Brass: Extruded Bar	2
E	Plinth	Cast Iron	1
F	Front Plate—Entablature	White Metal Die casting	1
G	Side Plate—Entablature	White Metal Die casting	2
H	Flywheel	Cast Iron	1
J	Cylinder	Gun Metal Casting	1
K	Cover—Cylinder	Gun Metal Casting	1
AA	Gland—Cylinder	Gun Metal Casting	1
AB	Steam Chest	Gun Metal Casting	1
AC	Gland—Steam Chest	Gun Metal Casting	1
AD	Cover—Steam Chest	Gun Metal Casting	1

Part Letter	Description	Material Supplied	No. of pieces
AE	Slide Valve	Brass Pressing	1
AF	Head—Valve Rod	Mild Steel $\frac{3}{16}$ in. Sq. $\times \frac{5}{8}$ in. long	1
AG	Valve Rod	Stainless Steel $\frac{3}{32}$ in. dia. $\times 2$ in. long	1
AH	Nut—Valve Rod	Brass $\frac{3}{16}$ in $\times \frac{3}{16}$ in. $\times \frac{1}{8}$ in.	1
AJ	Pin—Eccentric Rod	Mild Steel .172 in. Hex. $\times \frac{9}{16}$ in. long	1
AK	Sheave Eccentric	Cast Iron	1
BA	Strap—Eccentric	Gun Metal Casting	1
BB	Eccentric Rod	Mild Steel	1
BC	Stud—Eccentric	Mild Steel $\frac{1}{16}$ in. dia. $\times 1$ in. long	1
BD	Piston	Brass $\frac{3}{4}$ in. dia. $\times \frac{1}{2}$ in. long	1
BE	Rod—Piston	Stainless Steel $\frac{1}{8}$ in. dia. $\times 2\frac{1}{2}$ in. long	1
BF	Connecting Rod	Mild Steel $\frac{3}{16}$ in. $\times \frac{3}{8}$ in. $\times 3\frac{3}{4}$ in. long	1
BG	Bearings—Big End	Brass $\frac{3}{8}$ in. dia. $\times 1$ in. long	1
BH & BJ	Straps—Big & Small Ends	Mild Steel 20 SWG $\times 1\frac{1}{4}$ in. $\times \frac{3}{4}$ in.	1
BK	Cotters—Connecting Rod	Mild Steel $\frac{1}{16}$ in. $\times \frac{1}{2}$ in. $\times 2$ in. long	1
CA	Crankshaft	Mild Steel Fabrication	1
CC	Crankpin	Mild Steel $\frac{5}{16}$ in. dia. $\times 1$ in. long	1
CD	Crosshead Fork	Mild Steel $\frac{3}{8}$ in. sq. $\times 1$ in. long	1
CE	Slipper—Crosshead	Gun Metal Casting	2
CF	Bush—Crosshead	Brass $\frac{3}{16}$ in. dia. $\times \frac{5}{8}$ in. long	1
CH	Slide Bars	Mild Steel $\frac{3}{32}$ in. $\times \frac{1}{4}$ in. $\times 9\frac{5}{8}$ in. long	1
CJ	Spacer—Slide Bars	Mild Steel $\frac{3}{16}$ in. $\times \frac{5}{16}$ in. $\times 1\frac{1}{4}$ in. long	1
CK	Anchor Block—Slide Bars	Gun Metal Casting	2
DA	Flange—Exhaust	Brass Extrusion	1
DB	Pipe—Exhaust	Copper $\frac{3}{16}$ in. O/dia. $\times 18$ SWG $\times 3\frac{1}{2}$ in. long	1
—	Pipe—Steam	Copper $\frac{5}{32}$ in. O/dia. $\times 18$ SWG $\times 1$ in. long	1
DC	Flanges—Steam & Exhaust	Brass $\frac{5}{8}$ in. dia. $\times 1\frac{1}{2}$ in. long	2
DD	Drive Arms—Governor	Brass 22 SWG $\times 4$ in. $\times 2$ in sheet	1
DE	Carrier—Governor Weights	Gun Metal Casting	1
DF	Weight Arm	Brass 18 SWG $\times 2$ in. $\times \frac{1}{2}$ in. sheet	1
DG	Weight	Gun Metal Ball $\frac{1}{2}$ in. dia.	2
DH	Bracket—Governor	Gun Metal Casting	1
DJ	Nut—Governor Spindle	Brass $\frac{1}{4}$ in. dia. $\times \frac{7}{8}$ in. long	1
DK	Governor—Spindle	Mild Steel $\frac{3}{32}$ in. dia. $\times 3$ in. long	1
EA	Jockey Pulley	Brass $\frac{3}{8}$ in. dia. $\times \frac{3}{4}$ —n.	1
EB	Spindle—Jockey Pulleys	Silver Steel $\frac{3}{32}$ in. dia. $\times 1$ in. long	1
EC	Bracket—Jockey Pulleys	Brass $\frac{3}{32}$ in. $\times 1$ in. $\times 1$ in.	1
EC	Bracket—Jockey Pulleys	Brass $\frac{3}{16}$ in. dia. $\times 1$ in. long	1
ED	Pulley—Governor	Brass $\frac{5}{8}$ in. dia. $\times \frac{3}{8}$ in.	1
—	Spacing Bush—Governor Arm	Brass $\frac{3}{32}$ in. dia. $\times \frac{1}{2}$ in. long	1
—	Dowels—Column	Silver Steel $\frac{1}{8}$ in. dia. $\times 1$ in. long	1
—	Dowels—Entablature Panels	Brass $\frac{1}{16}$ in. dia. $\times 1\frac{1}{4}$ in. long	1
—	Pins—Governor	Brass 1 mm. dia. $\times \frac{3}{4}$ in. long	1
—	7BA Washers weight carrier	Brass	2
—	8BA Washers Jockey Pulley spindle	Brass	2
—	Oil Cup (Finished)	Brass $\frac{1}{8}$ in. dia. $\times 40$ TPI	2

List of Screws, Studs, Nuts, etc. Supplied to Build the 'Williamson' Engine

Description	Qty.	Material
$\frac{1}{2}$ in. × 6BA stud; column/entablature	4	M.S.
5BA × $\frac{5}{16}$ in. Hex. Hd setscrew; column/base	4	M.S.
8BA × $\frac{1}{4}$ in. Hex. Hd setscrews; base/exhaust flange	2	M.S.
7BA × $\frac{3}{8}$ in. stud; c/shaft bearing/entablature	2	M.S.
7BA × $\frac{1}{2}$ in. stud; c/shaft bearing/entablature/jockey pulley bracket	2	M.S.
8BA × $\frac{3}{8}$ in. stud; Governor bracket/entablature	2	M.S.
5BA × $\frac{3}{4}$ stud; base/plinth	4	M.S.
7BA × $\frac{3}{8}$ in. stud; cylinder/cover	4	M.S.
7BA × $\frac{7}{16}$ in. stud; cylinder/base	4	M.S.
7BA × $\frac{11}{16}$ in. Hex. Hd setscrew; cylinder/steam chest	2	M.S.
7BA × $\frac{9}{16}$ in. csk. setscrews; cylinder/steam chest	2	M.S.
7BA × $\frac{1}{2}$ in. stud; cylinder top cover/gland	2	M.S.
7BA × $\frac{5}{16}$ in. setscrew; slide bar anchor/cylinder cover	2	M.S.
8BA × $\frac{1}{2}$ in. stud; steam chest/gland	2	M.S.
7BA × $\frac{7}{8}$ in. bolt; Crosshead/connecting rod	1	M.S.
7BA × $\frac{1}{4}$ in. csk. setscrew; slide bars/anchorage	4	M.S.
7BA × $\frac{3}{4}$ in. bolt; slide bars/spacer	2	M.S.
7BA × $\frac{3}{8}$ in. Hex. Hd setscrew; fly wheel etc. (head to be filed square)	5	M.S.
6BA full nut	4	M.S.
7BA full nut	18	M.S.
7BA lock nut	3	M.S.
8BA full nut	4	M.S.
8BA lock nut	2	M.S.
5BA full nut	4	M.S.
10BA full nut	2	M.S.
10BA lock nut	2	M.S.
Washer, 7BA Governor weight carrier	2	Brass
Washer, 8BA Jockey pulley spindle	2	Brass
Not in Casting set—Order as Extra if required		
No. 155 Lubricator and Stopvalve Combined	1	(finished)
Lubricator Connector $\frac{1}{4}$ in. dia. × $\frac{1}{2}$ in. long	1	Brass